Trees of the Brain, Roots of the Mind

Giorgio A. Ascoli

The MIT Press
Cambridge, Massachusetts
London, England

© 2015 Massachusetts Institute of Technology

All rights reserved. No part of this book may be reproduced in any form by any electronic or mechanical means (including photocopying, recording, or information storage and retrieval) without permission in writing from the publisher.

MIT Press books may be purchased at special quantity discounts for business or sales promotional use. For information, please email special_sales@mitpress.mit.edu.

This book was set in Stone Sans Std and Stone Serif Std by Toppan Best-set Premedia Limited. Printed and bound in the United States of America.

Library of Congress Cataloging-in-Publication Data

Ascoli, Giorgio A.
Trees of the brain, roots of the mind / Giorgio A. Ascoli.
  pages cm
Includes bibliographical references and index.
ISBN 978-0-262-02898-1 (hardcover : alk. paper) 1. Neural circuitry. 2. Neural networks (Neurobiology) 3. Neural transmission. 4. Mind and body. 5. Brain. I. Title.
QP363.3.A83 2015
573.8′5—dc23
2014034243

10 9 8 7 6 5 4 3 2 1

Contents

Preface   vii
List of Figures   xiii

Part I: Cognitive Philosophy and Neuroscience Basics   1

1   Reality, the World, the Brain, and the Mind   3
1.1   What *Is*? Materialism, Idealism, and Dualism   3
1.2   The Brain, an Organ of Many   6
1.3   Cells of the Nervous System   9
1.4   The Conscious Mind   12

2   Neuron Trees and Network Forests   17
2.1   Neurons Connect into Networks with Tree-like Processes   17
2.2   The Roots of the Mind: Scales, Branches, Classes   20
2.3   Connecting the Dots: Synapses and Their Strength   24
2.4   Mind-Boggling Numbers of the Brain   27

3   Transmitting and Processing Information   33
3.1   The Axon: Signal Transmission   33
3.2   The Dendrites: Signal Integration   38
3.3   The Adaptive Neuron: Plasticity of the Input/Output Relation   43
3.4   The Mighty Cable of the Axon   49

Part II: Dynamics of the Brain-Mind Relationship   57

4   Activity Patterns and Mental States   59
4.1   The First Principle of the Brain-Mind Relationship   59
4.2   We Experience Only a Minute Fraction of Our Possible Mental States   63
4.3   Knowledge Is Encoded in the Synaptic Connectivity of the Network   66
4.4   A Far-from-Complete Engram   71

5 Learning from Experience 77
5.1 The Second Principle of the Brain-Mind Relationship 77
5.2 Probability versus Capability of Experiencing a Mental State 81
5.3 Panta Rhei 85
5.4 To Learn or Not to Learn 91

6 The Capability of Acquiring New Knowledge 95
6.1 The Third Principle of the Brain-Mind Relationship 95
6.2 Learning Is Gated by Background Information 101
6.3 Learning the Truth 108
6.4 Arbor Plasticity and Learning: Spatial and Temporal Scales 114

Part III: A Bucolic View of the Universe 123

7 Neurobotanical Gardens 125
7.1 Worms, Flies, and the Rest of Us 125
7.2 Arbors for All Seasons 132
7.3 Geographical Diversity and Physique du Role 137
7.4 Protagonists and Supporters 144

8 Neuron Types 153
8.1 All Neurons Are Different, but Some More Than Others 153
8.2 There's More Than Meets the Eye 161
8.3 Location, Location, Location! 165
8.4 A Multitude of Multitudes 171

9 Brain Branching and the Universe 181
9.1 The Liquid Jungle 181
9.2 The Uniqueness of the Individual 187
9.3 Brain Scans and Super-duper Microscopes 192
9.4 Neurofuturism: The Code Is the Beginning, Not the End 201

Epilogue 209
Notes 213
Index 229

Preface

BRAINFOREST GLORY: Since our first clear glimpse of a neuron over a century ago, humans have been awed by the beauty of these tiny tree-like objects. Perhaps our species' recent evolutionary stint as arboreal primates has hard-wired us to love the sights of trees and forests! Thus, as neuroscience rapidly advances our ability to see both the brain's "trees" and its vastly intricate and diverse forests, we may find that the most beautiful landscape of all is the one within.
—Stephen J. Smith

The human brain is often hailed as the most complex object in the universe. Much of the brain complexity is due to the massive web of connections and communication formed by its tens of billions of nerve cells through tiny tree-like structures. The computational power of such an intricate information-processing system is apparent in the depth and breadth of every sentient being's mental life. Although the complexity of the brain and the richness of the mind are nowadays readily recognized by scientists and nonscientists alike, my aim with this book is to reveal a yet well-kept secret: the stunning beauty of the brain's cellular form.

If we were to enlarge a human brain thousandsfold, to a first approximation each nerve cell would literally look like a tree with all of its intricate branching. A small region of the nervous system, such as those dedicated to moving your left pinky toe or to perceiving a dim blue light in the upper right corner of your eyesight, would at this magnified scale resemble a gigantic forest with millions of fantastic trees of all sizes and shapes. In this regard the entire brain can be viewed as a whole neurobotanical world completely filled with trees. These nerve trees are as beautiful and awe-inspiring as the most magnificent of magnolias, baobabs, and angel oaks. But unlike real-world trees, the microscopic trees in our brains make us move, feel, think, remember, plan, love, and enjoy life. I hope that my writing will inspire you to embrace them with your imagination.

The neurosciences are rapidly and increasingly entering into the mainstream discourse of our society, from economy to medicine, from technology to psychology, from politics to leisure. Yet media reports of new discoveries about the structure and function of the nervous system are by necessity somewhat detached from the complex technical details of the research laboratory. Even the most interested and educated professionals—lawyers, analysts, physicians, engineers—typically attain only a superficial understanding of the mechanisms actively investigated by neuroscientists. In contrast, the source information is described in specialized journals dense with jargon that is mostly unintelligible to outsiders. It takes on average more than a decade of dedicated doctoral and postdoctoral training for the best and brightest college graduates to develop the independent ability to access and contribute to the frontier of neuroscience knowledge.

Nevertheless, even the most experienced and accomplished researchers, when staring into their microscopes, are often captivated by the arresting splendor of nerve cells. It doesn't take a Ph.D. in neuroscience to enjoy the spectacular structure of the cellular constituents of nervous system any more than it takes a Ph.D. in music to appreciate Beethoven's Ninth Symphony. It seems unfair not to share the spring of such pleasure and inspiration.

Yet perhaps the biggest challenge to understanding our brains lies in connecting the subjective experience with the shape and activity of these nerve cells. We can measure the temperature of far-away galaxies with stupendous accuracy, but we are clueless when it comes to understanding why we remember certain episodes of our past and not others. This embarrassing ignorance creates a level playing field between the Nobel-winning neuroscientist and a random lumberjack. They both remember (or forget) what they had for breakfast the day before yesterday, and neither really understands the neural mechanism underlying her or his own mental state. Moreover, the neuroscientist and the lumberjack should both put their tools down from time to time and admire the scenic view of the forest in which they are working.

Somehow, that same structural majesty of nerve cells hides the secrets behind the genesis of our mental states, so immediately accessible to every human being and so elusive to scientific investigation. This book proposes new ideas to solve some of the most intriguing mysteries of the mind using only basic architectural principles of the brain. Why do we remember certain events of our past but not others? Why is it so difficult for me to learn certain skills although I can master others without effort, and yet it might be just the opposite for you? How do we acquire certain knowledge by reading or hearing something just once, whereas memorizing other facts might take multiple rehearsals? The novel explanations of these phenomena proposed in this book are consistent with the latest findings in the field but are not yet necessarily embraced by the entire scientific community.

In order to explain how beautiful tree-shaped cells relate to our brains' ability to store and even create new knowledge, we introduce and overview some basic mechanisms underlying the main functions of these nerve structures. Nevertheless, this book is emphatically *not* intended to provide a comprehensive introduction to neuroscience. Many other recent books are available on the brain at large, and we refer to several of them throughout our exploration as well as to excellent Web resources and, occasionally, to accessible reviews in the original scientific literature. These bibliographic references may serve as suggestions for recommended further readings or as initial pointers for interested readers to expand on specific topics.[1]

The book aims at a broad readership, generally employs plain language, and requires no specialized background knowledge. At the same time, some of the concepts I explain, both foundational and novel, are nontrivial. On top of it all I am firmly committed to scientific accuracy. I have strived to balance ease and pleasantness of read with a progressive buildup of logical steps by adopting an informal narrative style. When introducing each novel concept, I have attempted to include an explicit disclosure of whether it reflects established factual knowledge in the research community or represents commonly accepted but yet untested assumptions or constitutes a controversial view or even a completely original working hypothesis. To increase readability I have relegated the more technical explanations to footnotes for the benefit of the interested readers, but these are not required to follow the general flow of the arguments.

In the same spirit distances used to describe botanical trees are expressed in "imperial" units: miles (~1.61 kilometers), yards (~0.914 meters), feet (~0.30 meters), and inches (~25.4 millimeters). Microscopic measurements, however, typically referring to the subcellular constituents of neurons, are reported in metric units: nanometers, micrometers, and millimeters (one billionth, one millionth, and one thousandth of a meter, respectively).

The book is organized in three main parts. The first three chapters (part I) provide the scientific and logical foundations. Chapter 1 lays the relevant philosophical, biological, and cognitive grounds of the mind-brain relationship. Chapter 2 introduces the basics of nerve cells and how they interconnect to form a gigantic brain network. Chapter 3 explains the way nerve cells process, transmit, and store information by means of electric signals.

Certain sections of this first part (particularly section 3.2) may feel difficult to follow for some readers. If you find yourself fatigued or even lost through the details, do not despair. An in-depth understanding of those mechanisms is *not* necessary to grasp the key ideas of this book. Absorbing at least parts of the technical descriptions at the beginning, however, will enrich the appreciation of the subsequent material.

The following three chapters (part II) erect the central claim of the book by establishing a sequence of increasingly daring principles of the mind-brain relationship. Chapter 4 compellingly links mental states with patterns of electric activity in nerve cells. Chapter 5 presents an emerging minority opinion of how the brain adapts to learn from experience. Chapter 6 unveils a radically new hypothesis of the mechanism determining what is learned, what isn't, and why.

The last three chapters (part III) tie these concepts together with the cosmic diversity within and between brains. Chapter 7 describes functional differences and similarities of nerve circuits in disparate animal species, along the life span, and across brain regions. Chapter 8 provides a unified framework to embrace and comprehend brain complexity. Chapter 9 integrates these notions into a revealing perspective on the roots of individuality and humanity. Each of the nine chapters is subdivided into four thematic sections.

The ideas presented in this book arise from solid scientific foundations but lend themselves to tempting speculations about the future. Is there a hard limit to scientific knowledge? What is the ultimate human challenge? Although the answers must still be tentative, this is the first time in history that such momentous questions can be clearly formulated in neurobiological terms. Leaving the forecast of the fascinating world to come for the reader's contemplation, the book closes with an epilogue bringing us back to the present of contemporary neuroscience: what just happened, what's happening, what's in the making, and what lies as a challenge ahead.

The current and continuously accelerating pace of scientific progress makes for truly exciting times in research. It is hard to believe how much neuroscience has changed in the past generation, and it is even harder to imagine what our knowledge of the brain will be in the next generation. The most important task for neuroscientists today is to lay the foundation for the discoveries of tomorrow, but soon the findings of neuroscience will trigger profound changes in human relationships, transforming the very fabric of our society.

I wrote this book for my children, Benjamin, Ruben, Gabriel, and Jonah, for their enjoyment perhaps in a few years when some of these ideas might exhilarate them. I dedicate the book to my wife, Rebecca, with all my love: though the arbors of our brains are enclosed in different heads, the branches of our minds tenderly embrace.

This book would not exist without the direct or indirect contributions of several people to whom I am deeply grateful. First and foremost, I acknowledge my father, Aurelio, who taught me by example to search, research, and to experience life. It was during a 2009 vacation with him on mount Cervino in the Italian Alps that I conceived this book. Throughout the following five years, he continued to be a superhero grandpa and cared for my children for as many hours as I spent writing.

Bob Burke taught me to look at neuronal trees through the eyepieces of powerful microscopes. Steve Senft showed me the potential of computer simulations for neuroanatomy, and our early joint work planted the seed for many of the ideas in this book. Yuan Liu has been a visionary source of support through the years, venturing together in several expeditions through the neural forestry. Matteo Mainetti proved with numbers that the central and boldest proposal of this book (chapter 6) is mathematically tenable, but that's just one of the so many facets of our kinship. I celebrate two decades of friendship with Jim Olds, who provided much needed constructive criticism on an early version of the manuscript. Rebecca Goldin's extensive comments, incisive questions, and detailed suggestions, from the initial conception of this book through the penultimate draft, tremendously improved the readability of the final product.

My talented colleagues at the Krasnow Institute for Advanced Study year after year have provided an assiduous flow of ideas, passion, intellectual challenges, and personal warmth that always keeps me going. In particular, Harold Morowitz offered steady encouragement to pursue science off the beaten path and put me in touch with Bob Prior at MIT Press.

I'm honored to illustrate the beauty of trees through artful pictures by my awesome friend for over thirty years, Daniel Segrè (figures 1.1, 2.3, 3.3, 3.5, 3.6, 4.2, 5.1, 5.3, 5.5, 6.1, 6.2, 6.4, 6.6, 7.1, 7.6, 8.1, 8.3, 8.6, and 9.2). Our friendship started around campground fires surrounded by trees, and we have never lost touch with each other through our journeys across continents, scientific disciplines, and personal relationships. His photographs exude the boundless convergence of physics, biology, and human spirit. You really see his hand in them.

It is an inspiring privilege to work every day with such dedicated peers, students, technicians, and postdocs. I have certainly learned from my trainees at least as much as I was able to teach them, and I take tremendous pride from the shining success of my lab alumni all around the world. I especially thank Namra Ansari, Todd Gillette, Uzma Javed, and Amina Zafar, who skillfully illustrated neuronal data from NeuroMorpho.Org (figures 1.2, 2.1, 2.4, 3.1, 3.2, 3.7, 4.1, 4.3, 4.4, 5.2, 5.4, 6.3, 6.5, 7.2, 7.3, 7.4, 7.5, 8.2, 8.3, 8.4, 8.5, and 9.1). I am also grateful to George Mason University, Gerald Goldin, Michele Ferrante, and Evan Cantwell for figures 2.2, 2.6, 3.4, and 8.5, respectively.

The US National Institutes of Health, National Science Foundation, Office of Naval Research, Burroughs-Wellcome Trust, and Keck Foundation have been generously supporting my research over the years on neuronal trees and the NeuroMorpho.Org database, where the neurons gracing these pages (and tens of thousands more) can be downloaded.

Last but not least, the research behind the ideas proposed in this book is enabled by the courageous choice of many scientists to freely share their hard-won experimental data in publicly accessible databases, allowing reuse by their peers throughout the entire neuroscience community. Equally important, I am humbled by the selfless open-source software developers fueling the digital revolution. Thank you, and please know you are making a difference.

List of Figures

1.1 Neocortex   7
1.2 Concentrations and electric gradients   11
2.1 Soma, axons, and dendrites   19
2.2 November view from the author's office   21
2.3 One prototypical tree type to hold in your imagination   23
2.4 Axonal vines   25
2.5 Numbers of the brain   28
2.6 A didactic wander   31
3.1 Division of labor between axons and dendrites   37
3.2 Dendritic signal integration   41
3.3 Tree resembling neuron encoding tree   42
3.4 Dendritic branching   44
3.5 Active dendrites   48
3.6 The mighty cable of the axon   51
3.7 So tiny, yet so long   52
4.1 Place cells: neurons encoding space   62
4.2 Everything and the opposite of everything   64
4.3 Synaptic partners   70
4.4 Neural forestry   74
5.1 Learning from experience   79
5.2 Networking to learn   82
5.3 Mind the trees   88
5.4 In and out of the hippocampus   90
5.5 Branching arbors, memory's crucial ingredient   93
6.1 Their leaves might nearly touch each other   99
6.2 Wonder branching   103

6.3 The three principles of the brain-mind relationship   106
6.4 Convergence toward an idea   113
6.5 The structure-activity-plasticity relationship of the brain (and mind)   115
6.6 Enchanted forests   117
7.1 Hidden primate   127
7.2 Tiny or huge, squishy or crunchy, all multicellular animals (except sponges) have tree-shaped nerve cells   129
7.3 Changing brains in changing bodies   138
7.4 Actors and sensors   142
7.5 Projection neurons and local interneurons   148
7.6 Mood, trees, and neuromodulation   151
8.1 A botanic garden of shapes   154
8.2 Have axons, will travel   156
8.3 Chicken tree and chicken neurons   159
8.4 Beyond the appearance of neuron types   164
8.5 The neuronal circuit of the hippocampal formation: virtual and physical models   169
8.6 Homunculus   176
9.1 The far-reaching implication of axonal-dendritic overlap   183
9.2 Peering into the future   199

Part I: Cognitive Philosophy and Neuroscience Basics

1 Reality, the World, the Brain, and the Mind

... But of the Tree of Knowledge of good and evil you shall not eat of it, for on the day that you eat thereof, you shall surely die. (Gen. 2:17)[1]

And the serpent said to the woman: "You will surely not die. For God knows that on the day that you eat thereof, your eyes will be opened, and you will be as angels, knowing good and evil. And the woman saw that the tree was [...] desirable to make one wise; so she took of its fruit, and she ate, and she gave also to her husband with her, and he ate. And the eyes of both of them were opened, and they knew ... (Gen. 3:4–7)[2]

1.1 What *Is*? Materialism, Idealism, and Dualism

The material existence of the world around us is so obvious that we normally don't even bother questioning it. It requires, in fact, a certain mental effort *not* to take for granted the chair one is sitting on or the very ground holding the chair. Just think about this one book you are reading. You are reading it, so it surely *must* be there in your hands! And how about those hands of yours? You've known them, waved them, and used them all your life; they have always been there with you and for you. Of course they *are*.

Yet the questions of what is real and what is *reality* are neither trivial nor settled. We know that the chair, the ground, the book, and your hands have measurable physical properties such as weight and temperature. These measures can be determined reproducibly and objectively in the sense that they can be confirmed in different experiments and by independent experimenters. We also know that these objects are made out of molecules, whose chemical properties, also well known, can in fact explain the above macroscopic observables. Molecules are formed by atoms, atoms are formed by protons, electrons, and neutrons; and these particles are formed by the even tinier quarks. At these subatomic levels, however, it becomes impossible to determine precisely even simple measures such as position and velocity. Exactly what are the real constituents of matter, and what can be measured reproducibly and objectively?

To this day, these issues are the topic of advanced scientific research. Modern society made huge investments to build gigantic machines in order to enable the execution of complex experiments in the hope of finding answers. For example, construction of the 17-mile wide "Large Hadron Collider" particle accelerator by the European Organization for Nuclear Research (CERN) 500 feet underground in Switzerland between 1998 and 2008 had an approximate cost of $10 billion.[3] The project involved more than ten thousand engineers from over one hundred different nations and enabled the historical discovery of the long-theorized Higgs boson in 2012.[4] Data produced by this machine are distributed throughout a worldwide computing grid continuously analyzed by more than eight thousand scientists from 140 countries.[5] The theoretical framework necessary to describe and explain these issues employs mathematical formulations so sophisticated that they are elusive to all but a small number of highly educated specialists.

But even the so-called reality in the day-to-day life of people like you and me is not free of puzzling mysteries. Those chairs, books, and hands, as well as that ground we were considering are only "known" to us through the eyes of our minds. We perceive them through our sensations when we see them, touch them, and smell them. Even when we quantify their macroscopic physical attributes with dedicated instruments such as a scale, a thermometer, or what have you, we still rely on our senses to acquire those measurements. In fact, our whole experience of the external world is entirely enclosed in our mind. We do not "know" anything out there *directly*.

What is even more unsettling is that our perceptions have been shown over and over to be fallacious. Thousands of well-known optical illusions can fool our sight, convincing us that perfectly static images are moving, that nearby objects are far away, and that two identical shapes are different. We also routinely hear sounds that are not there, we confuse tastes, and misjudge textures. We detect patterns when there are none, and we miss most of what surrounds us. And who has never witnessed the most vivid, real, and unquestionable experience, only to wake up and realize it was just a dream? Might it not be *all* a dream?[6]

Let's face it: the deeply rooted presumption that we are embedded in a material world to start with, let alone that the world is somewhat close to its appearance, may after all be nothing more than a belief. Accepting this supposition constitutes a logical leap of faith, in mathematical terms an *axiom*. It constitutes in fact the cornerstone of the philosophical theory of *materialism*.

There are, however, possible alternatives. If our entire world perception is at least mediated by the mind, perhaps it would be simpler to assume the mind as a starting point and postulate that the appearance of matter is a product of the mind. The strict logic of this argument constitutes the foundation of *idealism* and has met the favor of major philosophers throughout the centuries, from Plato to Kant and Hegel.

Yet, to many, idealism is not fully satisfying either. First, if our mind creates the world, how about the "other" fellow beings? If each of them is also creating a world on his or her own, from a behavioral point of view it seems awfully similar to ours. We agree on too many details of our surroundings to maintain that they are independent creations of distinct individual minds. Second, it may be troubling to equate the reality of concrete objects, such as a chair or a hand, to that of abstract concepts or inner feelings, such as probability and love. They all might be a product of the mind, but they surely *feel* very different.

A third alternative to materialism and idealism is to consider both matter and mind as fundamental constituents of reality. This line of thought is known as *dualism*. In Descartes's classic formulation, dualism purports matter and mind to be irreducibly distinct substances that nonetheless causally interact: physical events cause mental events, and mental events cause physical events. Although this position accounts for both the material world and mental phenomena, it does not provide a direct solution to the ultimate philosophical problem of the mind-matter relationship. Instead, it shifts the issue to the question of their *interaction*. Certain organized forms of matter, most notably nervous systems in specific dynamical states, appear to be systematically associated with feelings, thoughts, memories, and intentions. But why is this so? Logically, it is easy to imagine a material world identical to ours in all respects, with human beings and their behavior as we know it, but with no meaning and inner life.

Humankind has pondered the question of what actually "is" for as long as we have written records, and this issue is still debated.[7] In George

Berkeley's famous words, "What is mind? No matter. What is matter? Never mind."

Will there ever be a scientific solution to this problem? All scientific theories ultimately require axioms at their foundations. With enough experimental evidence, all phenomena might eventually be precisely quantified and described within the framework of rigorous mathematics. If we capture the essence of the relevant material structures and mental phenomena in a way amenable to a satisfactory and predictive mathematical formalism, it might be that in the end the formulas describing matter and mind will be one and the same.[8] In this case there will be a set of equations stating the correspondence between the two. By the same token, we already know that electricity *is* shifting charge, optics *is* electromagnetic waves, temperature *is* movement of molecules, and genetics *is* interaction of nucleic acids. Perhaps one day we will have a similar understanding of matter and mind.[9]

1.2 The Brain, an Organ of Many

Our bodies and lives depend on many organs: kidneys, liver, lungs, heart, blood, skin, stomach, intestine, ... and the brain. To be more precise, there are a central and a peripheral nervous system. The *central nervous system* (defined as the part of the nervous system surrounded by bones) consists of the brain and the spinal cord. There is massive empirical evidence that mental phenomena are associated with brain activity. Even emotional feelings and intuition, which common language still refers to as originating in the heart and the gut, are uncontroversially rooted in the brain. Yet the brain is far less homogeneous than all other organs. It is composed of so many discrete parts with such disparate functions that it is perhaps more useful to consider the brain as an ensemble of different suborgans. Although intimately interacting together, these components are physically and functionally distinct.

When we think of a human brain, the popular image of the wavy convolution of folds that comes to mind corresponds to the cerebral cortex or *neocortex*, the outer structure that covers almost the entirety of the brain surface. Cortex is the Latin term for "bark," evoking the external texture of the tissue overlaying woody plants (figure 1.1).

Figure 1.1
Neocortex (Daniel Segrè, 2014, Idaho).

The cerebral cortex is involved in many cognitive functions and can be coarsely divided into main geographic areas. The region at the back, called the *occipital lobe*, is associated with vision. The region at the top, called the *parietal lobe*, is associated with somatosensory perception (touch) throughout the body. The regions on the (right and left) sides, called the *temporal lobes*, are associated with hearing and imagery of objects and faces. The anterior region, called the *frontal lobe*, is associated with motor control, decision making, planning, and abstract thought. As does most of the rest of the brain and body, the cortex has a bilateral organization. The two hemispheres are connected by a thick bundle of nerves called the *corpus callosum* (Latin for "corny body").

The cerebral cortex communicates with the rest of the nervous system through a structure that lies underneath the surface, called the *thalamus* (Greek for "chamber"). The thalamus relays signals to and from the suborgans responsible for producing and executing the information processed in the cortex, including the retina in the eye, the cochlea in the ear, and the spinal cord, which contacts muscles and skin receptors. Below the thalamus lies the *hypothalamus*, which controls the release of hormones regulating hunger, thirst, mood, and body temperature. Next to the thalamus stands the *basal ganglia* (Latin for "at the base" and Greek for "knots"), which is associated with controlling motor strength, habitual behavior, task automation, and addiction.

At the back of the brain rests the *cerebellum* (Latin for "small brain"). The cerebellum contains more nerve cells than all the rest of the brain together and is responsible for coordination of activity, fine movements, and adaptation to unexpected events. Between the brain and the spinal cord we find the *brainstem*, which is responsible for heartbeat, respiration, and circadian rhythms including sleep and dreams. Somewhat wrapped in this mass of tissue hides the *amygdala* (Greek for "almond"), a structure involved with emotional learning and fear. The *hippocampus* (Latin for "seahorse," after its shape) has the complementary roles of consolidating and retrieving autobiographic memories and intentions with a "first-person" (egocentric) perspective as well as mapping locations and context.

The above brief overview is far from comprehensive and grossly oversimplified. There are many other suborgans in the brain, and each of these regions has a complex internal organization with functionally specialized fine substructures. Nonetheless, it should already be apparent that the nervous system performs many more functions than any other organ. Indeed, the complex composition of the brain reflects the diversity of its functions. Yet much of the stunning brain complexity is due to its *cellular* architecture. In spite of their distinct positions, shapes, and roles, all parts of the brain share common organizational principles and are assembled out of similar building blocks. The mind-matter relationship must ultimately be explained in terms of the structure, activity, and unique properties of these cells.

1.3 Cells of the Nervous System

The brain volume is occupied by nerve cells, also called neurons, as well as by glial cells (often referred to as "glia") and blood vessels (veins, arteries, and capillaries). Only the former two, neurons and glia, technically belong to the nervous system, as the cellular constituents of the vasculature (erythrocytes, endothelial cells, etc.) are contributed by different organs. *Glial cells* (from the Greek "glue") play many fundamental support roles. They form a scaffold around the membrane of neurons, facilitating signal propagation. They aid messaging at the connections between neurons. They mediate the interaction with the bloodstream, filtering chemical intake, providing neurons with highly selected nutrients, and removing toxins and metabolic waste. They also destroy pathogens, remove dead neurons, and guide new neurons during development. *Neurons*, however, are the unquestionable protagonists in the representation, processing, transmission, and storage of information. This book is about *them*.

Neurons carry out their action by turning on and off. However, neurons act collectively, each playing an individual role in a gigantic symphonic orchestra. As I explain later in the book, the concerted spatial and temporal activity patterns of large collections of neurons underlie all mental states. In other words it isn't the activity of a single neuron that represents the chair, or the ground, or the book, or the hands. Instead, each mental state is represented by an ensemble activity distributed over many neurons. Such a "population code" enables a much greater repertoire of states, as well as robustness to noise. What the Greek philosopher Aristotle said in the *Politics* referring to human individuals and the value of democracy very much applies to neuronal organization: "For the many, when they meet together, may very likely be better than the few good. Hence, the many are better judges than a single one, for some understand one part, some another, and among them they understand the whole."[10]

How do neurons turn on and off? Inside their membrane neurons actively maintain concentrations of certain chemicals that are different from those found outside, in the extracellular space. This inequality is often referred to as a *gradient*. Specifically, there are more sodium and chloride ions (the components of common sea salt) outside of neurons than inside. The term *ion* refers to particles that carry an electric charge: sodium is charged positively, and chloride is charged negatively. Moreover, there are more (positive) potassium ions and various types of other (negative) ions inside of neurons than outside (figure 1.2). The balance works out such that, in their typical state, neurons are negatively charged inside their membranes relative to the extracellular space. Thus, there are both a *concentration gradient* for certain ions across the membrane and an *electric gradient* due to a charge imbalance.

When a neuron turns on, its electrochemical gradient gets temporarily shorted and quickly restored, similar to the momentary flicker of a light

Figure 1.2
Concentrations and electric gradients. (Top) When a neuron is "resting" in its typical state, chemicals are not distributed evenly across the two sides of its cellular membrane. Sodium (positively charged) and chloride ions (negatively charged), the components of common sea salt, are much less concentrated inside than outside, indicated in the figure by the relative font sizes. In contrast, potassium (positively charged) and other negatively charged substances (anions) are more concentrated inside than outside. As a consequence of this uneven distribution, called a *concentration gradient*, the inside of the membrane is negatively charged relative to the outside (indicated by the minus and plus signs), an imbalance termed *electric gradient*. (Bottom) Graphic rendering of a neuronal electric pulse. These two images represent an interneuron from the outer granular layer of the rat olfactory bulb.[11] The cell body (the "center" of the neuron) is the very small spherical shape close to the base of the tree-like structure. In this neuron type the arbor around the cell body receives and processes the inputs from other neurons. The arbor rising to the top carries the output signals to other neurons. On the left the input tree and cell body are colored in shades of red, and the output tree is in a gradient of brown (rendering by Uzma Javed in the author's lab). On the right the colors are changed (by the author) to illustrate the temporary voiding of electric gradients caused by sodium influx during rapid neuronal activation. This reconstruction is freely available online at NeuroMorpho.org (branch thickness was increased to enhance contrast).

bulb or computer monitor. In particular, in very rapid sequence, sodium first flows in, voiding the electric gradient as well as its own concentration gradient. Next, potassium flows out, restoring the electric gradient but voiding its concentration gradient. Last, a molecular exchange machinery pumps sodium out and potassium in, restoring the chemical gradient. This series of events lasts only a couple of milliseconds (thousandths of a second) back to back.

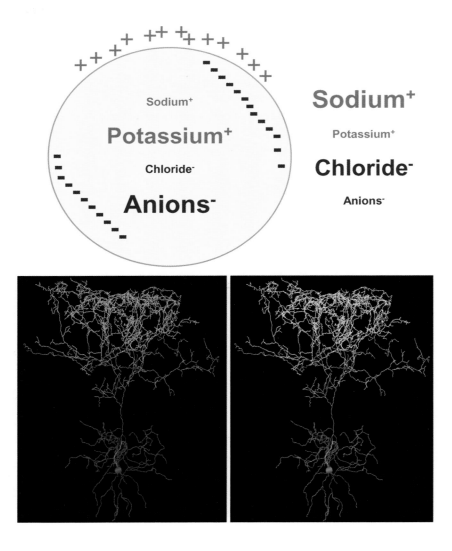

Neurons are able to use this rapid electric pulse to communicate with each other. Specifically, they exchange information by means of elaborate trees that can extend across the entire brain! We introduce these very special arbors in the next chapter. Then, in chapter 3, we reveal several additional aspects of the mechanisms of neuronal activation and of information processing, transmission, and storage. In later chapters we also explore the diversity of neurons in different animal species, during development and aging, across brain regions, and in relation to their functional specializations. First, however, we turn our attention to consciousness itself.

1.4 The Conscious Mind

The nervous system performs many vital functions, including control of respiration, heartbeat, blood pressure, hormonal release, digestion, the sleep-wake cycle, and reflexes. From the survival point of view, consciousness does not even make the top-ten list! Indeed, from the perspective of scientific materialism, there is no need for consciousness at all. We could logically imagine a world of human beings with their brains, bodies, and behaviors, even conversations, but absolutely no inner experience.[12] Yet we do have inner experience, and in fact it is precisely that inner experience that makes our lives meaningful and interesting. Of course we *need* the brain to keep us alive, but we *care* about our memories, emotions, intentions, plans, feelings, thoughts, and reasons. These mental states have proved to be more difficult to characterize and quantify objectively than the physical properties of a chair or a book. However, we immediately *know* our mental states more intimately and directly than any physical object out there. By this measure, mental states are certainly observable: they are the most common observations of all!

If we want science to explain and predict as many observables and observations as possible, then we need to characterize the relationship between the brain and the conscious mind.[13] We cannot simply accept the adage that the "conscious mind is what the brain does" because the brain does many things that are not mind. And we cannot revert to the purely phenomenological view of the conscious mind as "what it is like to be" because there is no part of the brain in it, which makes any scientific breakthrough about the underlying mechanisms unlikely. Instead, we propose a definition to bridge these two aspects:

The conscious mind is what the types of brain activity that feel like something feel like.

Admittedly, such a mouthful is not easy to parse. Let's take a little time to understand what it means.

We start from our feelings as the "given" observable aspects of mental states, in the sense that they can be observed by ourselves. We note that these feelings constitute a quintessential aspect of brain activity—but not all or any brain activity. Only some types of brain activity are linked to feelings. Then we consider those specific types of brain activity (the ones linked to feelings), and we ask what they feel like. Thus, we can equate the conscious mind to what the relevant types of brain activity feel like. To aid logical parsing, the same definition can be broken by parenthetical separations:

The conscious mind is what {the types of brain activity [that feel like something]} feel like.

By bringing together corresponding brain and mental states, this definition may prove to be useful for exploring their relationship.

An interesting property of mental states is their duration. Different percepts, thoughts, and emotions can last for varying spans of time, from a fleeting visual flash to an extended meditative contemplation. The extent of an uninterrupted mental state might depend only on our ability to concentrate and the stability of the surrounding environment. We cannot apparently experience, however, arbitrarily short mental states. There seems to be a minimum duration requirement of approximately 50 milliseconds for any inner content to be consciously accessible. This means that we can at most experience some twenty or so different mental states per wake second of life. The average life expectancy in the industrialized world amounts to approximately 34 billion moments of consciousness per human being.

Shorter events can still affect our behavior, but only subliminally. This minimal "time unit" of the conscious mind varies slightly by individual and by modality (auditory vs. visual inputs, motor commands vs. tactile sensation, etc.) and is also affected by fatigue and distractors. Subliminal suggestions have been used in a variety of applications from advertising tactics to self-help techniques for confidence boosting. For example, a single frame depicting a particular brand of soda inserted in a webcasted sports event might increase your craving for that beverage during and after the game. However, you would remain completely oblivious to the fact that the suggestion was implanted in your mind. Such a trick might be considered deceptive and morally objectionable, and subliminal advertising is banned in several countries such as Great Britain and Australia, but not in the United States.

A fundamental aspect of the conscious mind is the balance between stability and plasticity. To immediately experiment with mental stability, try this simple exercise. Focus for a few seconds on the first letter of this page, until you gain a firm sense of the mental state. Then put the book down and take a little break. Walk a few steps, clap your hands, think about yesterday, and tomorrow. Then, after experiencing those other mental states, pick the book back and focus again on that same first letter of this page. Most likely, it will *feel* very similar to the first time around. You might think that's because the *letter* didn't change. Its shape, position, color, and other properties remained the same; isn't that a good enough reason for our conscious experience of it to do the same? The short answer is *"No!"*

On the one hand, a stable physical signal does not imply unchanging conscious experience. Let's say that when opening the newspaper you notice a photograph of a candidate for the local election who looks familiar. You stop and stare at the image for several seconds, and suddenly it dawns on you that the politician featured in the paper is actually a former grade school classmate of yours. The face in the picture didn't change any more than the first letter on this page, yet your mental state transitioned sharply from a "tip-of-the-tongue" experience to the firm conviction of a childhood memory. On the other hand, a steady stimulus is not required for ensuring the robustness of a percept.

Let's say tonight, just before falling asleep, you resolve to be more patient with your colleagues at work. You sleep an uninterrupted night, alternating several times completely mindless black-ins with vivid dreams (only the last one of which you might remember when you wake up). In the morning you get out of bed and go about your routine. Chances are, you still have the intention in mind to keep your cool with the co-workers. Whether you succeed or lose your temper later that day, the mental state of the intent remained stable since the previous day in spite of a restfully resetting sleep cycle. Yet there's nothing physical or objective in your resolve, you didn't tell anyone, and there's no unchanging letter in the corner of the page to ground your mental state.

A seemingly contrasting aspect of the conscious mind is that it also continuously changes. As we go about our lives, we experience events and mental states, and those experiences alter what and how we'll perceive later. In the previous example, before resolving to be patient with your colleagues, that mental state was not part of your inner life. Having had the thought modified your feelings the next day. The ability of our brains to change thought patterns is called "plasticity." Although our mind does adapt to the environment, it is not necessary for an external event to occur to trigger the plasticity of the self. Sometimes a single occurrence of a particular mental state can dramatically and irreversibly change the rest of our existence.

Obviously not *all* mental states have the same power to alter our minds. Most of our experiences are fleeting and leave little if any trace. Indeed, it would be disastrous if we could remember every single mental state we have ever experienced![14] What determines the strength by which the mind is affected by different mental states? We seem to have an uncanny ability to select the knowledge we acquire by how useful or *predictive* it is. If a bus stops at the corner while a kid calls his dog, we might associate the bus number with the location of the stop and the dog with its name and owner. Somehow we are less likely to associate the name of the dog with the bus number or the kid's face with the location of the bus stop. Why is that?

More generally, why do certain experiences make sense and others don't? It seems clear that background knowledge plays an essential role here. Experienced piano players can read music scores and later even remember the tune in their minds without having heard or played the piece. People who never played piano can look at the same music score, see the same notes in the same position, but they will not "hear" the tune, nor will they remember it the next day.

If we can explain these observations in terms of neuronal mechanisms, we will begin to solve the mind-brain problem.

2 Neuron Trees and Network Forests

2.1 Neurons Connect into Networks with Tree-like Processes

Among the various types of cells within the brain, the previous chapter introduced the neurons as the principal actors on the stage of the mind. Neurons are among the most peculiar cell types across all of life's kingdoms. Heart cells, kidney cells, liver cells, blood cells, and lung cells all perform vital functions by means of impressive biochemical machineries. Yet it is the activity of neurons that somehow gives rise to our inner lives. What makes neurons so special? In chapter 1 we saw that resting neurons maintain an electrical and chemical gradient between inside and outside of their membranes. When active, neurons can temporarily level out or even invert this gradient, a capacity that makes them *excitable cells*. It turns out that other cells in our body share this property of being excitable, including muscle, glial, and endocrine cells (those releasing hormones into the bloodstream). Why do we reason with the brain and not with biceps and glands?[1]

The prototypical biological cell is illustrated in many textbooks as a more or less spherical shape, with an external membrane enclosing a watery mix of internal organelles such as mitochondria (the powerhouses that burn sugars to generate energy), ribosomes (the assembly lines that produce proteins, the ubiquitous macromolecular machinery), the endoplasmic reticulum (where most chemical reactions take place), and the nucleus (the control center that contains DNA and dictates genetic expression). Every neuron has all of that in its cell body, which is also called the *soma* (Greek for "body"). The soma (plural: *somata* or somas), however, constitutes only a minute part of a neuron. The largest portion consists of extensive arbors that originate from the soma and extend out for considerable distances. These tree-like processes enable distant neurons to connect with each other.

That is what makes neurons so special in the relation between matter and mind.

There are two kinds of neuronal arborizations, called *dendrites* and *axons*. To a first approximation, dendrites constitute the informational input of a neuron, whereas axons constitute its informational output. In other words a neuron receives signals from other neurons on its dendrites. These signals travel to the soma, and the soma sends its messages out to other neurons through its axon. Most neurons have only one axonal tree and one or more (up to a dozen, but usually fewer) dendritic trees (figure 2.1). The next chapter offers more details of signal processing in dendrites and axons. For the time being suffice it to say that both axons and dendrites branch profusely. A typical neuron has anywhere between a dozen and many thousand branches.

Dendrites and especially axons extend their branches through a volume that is much, much larger than the soma they are attached to. The tips of the dendrites often terminate at a distance that is hundreds of times greater than the diameter of the soma. The axons can span an even larger space. To cite an extreme case, there are neurons whose individual axons travel continuously from the top of our head to the bottom of our back. The axons of another group of neurons go from the bottom of our back to the tip of the toe. Thus, just two axons are sufficient to span the entire length of our body (for professional basketball players, that's almost 7 feet!). In striking contrast, the soma is a microscopic dot, one-tenth of the tip of a needle! If one

Figure 2.1
Soma, axons, and dendrites. (Top) A layer 2/3 interneuron from the rat somatosensory cortex.[2] The soma and dendrites are colored purple, axon green; the left and right insets zoom in on the arbors near the soma and on more distal branches, respectively (rendering by Uzma Javed in the author's lab). (Middle) A pyramidal neuron from the mouse somatosensory cortex.[3] The soma, the apical trunk of the dendritic arborization, and an incomplete local subset of the axonal arbors are colored yellow; the rest of the dendritic tree is brown; the left and right insets zoom in on neuronal branching and the somatic region, respectively (rendering by Amina Zafar in the author's lab). The much more extensive distal axonal branches were not reconstructed. (Bottom) A basket cell from the mouse neocortex.[4] The soma and dendrites are green, axon is blue; the inset zooms in on the somatic region (rendering by Uzma Javed in the author's lab). These reconstructions are freely available online at NeuroMorpho.org (branch thickness was increased to enhance contrast).

were to look at the entire span of a single axonal arbor, the soma in its midst would be so tiny as to appear in fact invisible.

Such extensive arbors enable each neuron to reach out to tens of thousands of other neurons, sometimes even 100,000. For comparison, there are about 25,000 trees in Central Park[5] of Manhattan, NY. Try to imagine an

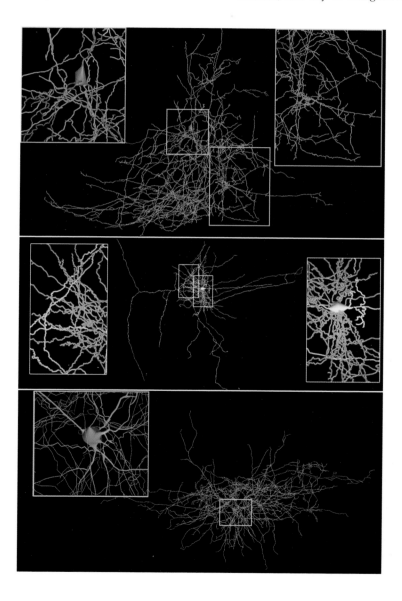

arbor that can touch one leaf of each and every tree[6] in Central Park. That's quite a feat already. Now consider that each of these trees has some 25,000 leaves,[7] and each of those is touched by one such gigantic arbor. What you have is already a massive web, a *network* of 25,000 × 25,000 (more than half a billion) contacts. And this is just to account for the contacts received by all the neurons contacted by a single neuron! The brain has many, many neurons, and the gargantuan network they form is so huge as to defy our ability to visualize it in our mind. This is one of the reasons behind the growing utility and requirement of computational techniques to analyze and synthesize neuroanatomical data.[8] In order to help us comprehend the complexity of brain circuitry, this book continuously refers to the botanical metaphor to explore a bit more of these (literally!) mind-boggling neuronal trees, their interconnections, and the sheer inner size of the brain.

2.2 The Roots of the Mind: Scales, Branches, Classes

Although the axons of some neurons can extend for several feet, neuronal dendrites are more contained. In the typical neuron the dendritic tree span is approximately 1–2 millimeters, just about the thickness of a penny. Imagine taking this really tiny tree and enlarging it around one thousand times, so that it would be about your height; that's a little more than what neuroscientists do when they look at neurons through light microscopes. What would this human-size tree look like? The most striking aspect would certainly be how slender its branches are. Even at its thickest, close to the base, the trunk would be just as thin as a toothpick! If you tried to "shake hands" with one of the side branches of this dendrite (expanded to be as tall as a person), you would be clasping a limb no thicker than a hair!

After this close encounter let's mentally dilate this dendrite another ten times, so as to make its vertical elevation similar to that of a typical medium-size tree.[9] At 60 feet high this magnified dendritic tree would nonetheless have a trunk about the thickness of your pinky finger (hardly something you could hug, let alone climb on!), and its side branches would be as thin as spaghetti. Why are neuronal trees (relatively) slimmer than woody trees? The reason for this difference is that botanical trees have to sustain their weight in the face of gravity. Although branches of nearby botanical trees can cross each other, their trunks, even deep in the woods, are typically

Figure 2.2
November view from the author's office (Fairfax, Virginia, 2007. Photograph courtesy of George Mason University).

dozens of feet apart. The vast majority of the surface of a botanical tree touches nothing but air. Neuronal trees, in contrast, are jammed together to completely pack the surrounding space. Similarly to a bus ride in peak-commute time, when passengers become sardines and could not fall on the floor even if they stopped standing on their feet, neurons have no gravity to fight, just space to fill up. This is why their branches can afford a slim design, which enables greater length and higher packing density within the same brain volume.

Aside from this difference, neuronal dendrites really look like miniaturized trees. The similarity is so striking that when I first started in a neuroscience lab, every time I looked at neurons through a microscope, I saw just a bunch of trees. Now, after thinking every day for twenty-five years of the mind-brain connection, when I look out of my office window, I see neurons (figure 2.2).

No two biological objects are exactly alike, and that holds true for both botanical trees and neurons. One can walk for miles in an oak tree forest without ever encountering two trees with all the same branches in the same positions. Yet two oak trees are clearly more similar to each other than an oak tree to a pine tree. It is just the same for neurons. There are many types of neurons. They have names such as pyramidal cells, granule cells, basket cells, chandelier cells, and ganglion cells, but they might just as well be called oak trees, pine trees, cedar trees, willow trees, and palm trees. We explore more of the differences among neuron types in chapters 7 and 8, after we discuss possible mechanisms by which their arbor shapes relate to mental phenomena. For the time being, suffice it to say that the diversity of botanical shapes is an excellent metaphor for their neuronal counterparts.[10] Some trees are smaller; others are bigger. Some have branches coming out of the trunk at acute angles; others are nearly perpendicular. Still other trees do not have a trunk at all, sprouting all of their branches like a shrub. Some trees have straight branches, others meandrous, and others yet climb like vines. Some trees can have hundreds of short branches and others just a dozen longer ones. Some trees have smooth bark; others have thorny protrusions. All of these descriptors and many more are in fact perfectly applicable to dendritic and axonal arbors.

Keeping in mind such stunning diversity, pick one prototypical tree type to hold in your imagination (figure 2.3). Let's consider how neuronal trees communicate with each other.

Figure 2.3
One prototypical tree type to hold in your imagination (Daniel Segrè, 2014, Fresh Pond Park, Cambridge, Massachusetts).

2.3 Connecting the Dots: Synapses and Their Strength

Unlike botanical trees, neurons do not operate in isolation. On the contrary, their entire reason for being is to communicate with each other. In fact their tree shapes are the *means* by which this communication occurs. Axons are the neuron loudspeakers: they send the signal out to the tens of thousands of other neurons they can reach. Dendrites are the neuron receivers: they listen to and integrate the incoming messages from (the axons of) tens of thousands of other neurons. The point of contact between the axon of a neuron and the dendrite of another neuron is called a *synapse*.[11]

Relative to the synapses they contribute to, the axon and the entire neuron it belongs to are referred to as "presynaptic," whereas the dendrite and the neuron it belongs to are "postsynaptic." Each neuron is presynaptic to tens of thousands of synapses and postsynaptic to tens of thousands of other synapses. That is to say, each neuron has tens of thousands of inputs and tens of thousands of outputs. Although dendrites and especially axons often form synapses at their terminal tips, the vast majority of synapses are formed along the axonal and dendritic branches. Axons typically swell into small beads, called *varicosities*, corresponding to the synapses they form (figure 2.4). Dendrites can receive synapses on small protrusions called *spines* or directly on the branch shaft. The tiny space in between the axonal varicosity and the dendritic spine or shaft is called the *synaptic cleft*.

Axons and dendrites are *electrical* devices. Their signals consist of small pulses of electricity that travel from the soma down the axons and from the dendrites toward the soma (we see a few more details on this in the next chapter). However, electrical signals do not themselves cross synapses.[13] Instead, synaptic communication occurs by means of *chemical* transmission. When the pulse of electricity coming through the axon of the presynaptic neuron reaches each of the axonal varicosities, it causes the ejection of some molecules, called neurotransmitters, into the synaptic cleft. When they are recognized by the dendritic spine or shaft receptors, neurotransmitters trigger a new electrical signal in the dendrite, which starts propagating toward the soma of the postsynaptic neuron.

When I first learned of this byzantine mechanism I could not help asking: Why doesn't the electric pulse pass straight from the axon to the dendrite? It would be much faster, certainly simpler, and more reliable from

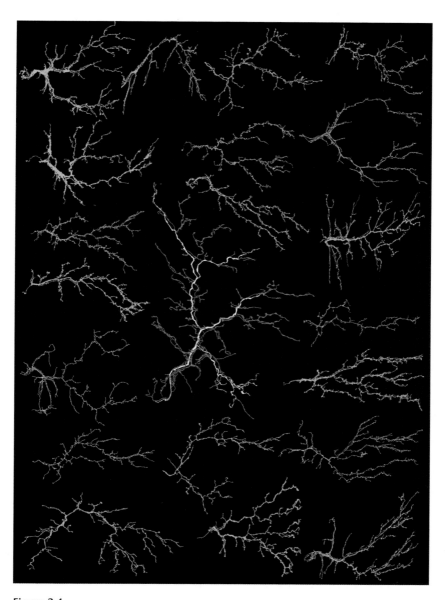

Figure 2.4
Axonal vines. Terminal arbors of twenty-one climbing fibers in the rat cerebellum (data by the author[12]) with the center image enlarged for detail and colored with the main trunk white and the side branches in shades of red (center image rendering by Amina Zafar in the author's lab). All reconstructions are freely available online at NeuroMorpho.org (branch thickness was increased to enhance contrast).

pulse to pulse. It turns out that replacing chemical transmission with the direct passage of electricity from axons to dendrites in our brain would likely annihilate most of cognition as we know it! The reason is that molecular neurotransmission, or chemical communication, provides a great degree of flexibility in the type and amount of information flow between neurons. This flexibility is quintessential to how the brain works.

First of all, although the axonal signal is pretty much the same from time to time and in all neurons, different neuron types release different neurotransmitter molecules. As we see in chapter 3, different neurotransmitters have dramatically different effects postsynaptically. Some neurotransmitters are *excitatory*, which is to mean that the electric pulse they trigger in the dendritic spine or shaft encourages the start of an axonal pulse in the postsynaptic neuron. Others are *inhibitory*, having literally the opposite sign of the excitatory signals and therefore making the postsynaptic neuron less likely to start its own axonal pulse. Yet other neurotransmitters are *modulatory*, in that they affect how sensitively the postsynaptic neuron should respond to incoming excitatory and inhibitory signals without weighing in directly on the decision whether to send a message down its axon. Every neuron receives a combination of excitatory, inhibitory, and modulatory contacts. This constant polling of information is not unlike a referendum, whereas excitatory and inhibitory signals correspond to "yes" and "no" votes, and modulatory signals set the quorum for a decision.

The second and perhaps most essential feature enabled by chemical transmission is that it allows individual synapses to fine-tune the *strength* of their communication. The same axonal pulse might trigger a stronger dendritic signal in one neuron and a weaker one in another. Most importantly, the same synapse can change its strength from time to time, a phenomenon known as *synaptic plasticity*.

There are both presynaptic and postsynaptic mechanisms in place to enable such flexibility and plasticity. For example, an axonal varicosity can release more or fewer neurotransmitter molecules into the synaptic cleft. Moreover, each neurotransmitter molecule can trigger a larger or smaller signal into the dendrite, and the duration of this postsynaptic signal can be changed too. Thus, every neuron can independently "decide" the strength of each of the tens of thousands of signals it sends to other neurons and of each of the tens of thousands of signals it receives from other neurons. In the "political metaphor" of synaptic polling as a referendum, this adaptive nature of neuronal signals means that each vote does *not* count the same! Some votes are heavier than others, but this differential weighting scheme changes continuously. Thus, if you were a synapse, you might have a bigger say in the coming election but become disenfranchised in the one that follows.

The exquisite ability of neurons to adjust their synaptic strength has long been thought to underlie the uncanny capacity of brains to learn from experience.[14] In chapters 4 and 5 we argue that the story may be a bit different from this popular view. The alternative neural mechanisms we propose there are more consistent with the most recent research breakthroughs and provide a tighter logical connection between the brain and the mind. These "new" mechanisms also offer compelling explanations for a broader class of cognitive phenomena.

2.4 Mind-Boggling Numbers of the Brain

There are somewhat shy of 100 billion neurons in a human brain,[15] a number hard to imagine. For comparison, take a look at figure 2.5, which may initially seem like an innocent gray canvas. If you stare closely at the dull-looking page or screen, bringing it near enough to your eyes, you might be able to make up the individual points of ink (black) or space (white). There are some 175,000 such points in this figure.[16] Even if this whole book were entirely made of those "gray" pages, it would still only contain fewer than 40 million dots, the number of neurons in less than a single cubic centimeter of neocortex! To simply plot a dot for each neuron of a single human brain, it would take over 2500 copies of this book (but then it might be hard to find as many buyers for a 200-page book made up entirely of miniaturized checkerboards).

Every neuron connects on average to 50,000 other neurons. Printing all of the synapses of your own brain a dot at a time would fill all the pages of 100 million copies of this book, equivalent to approximately three times the entire collection of the US Library of Congress[17] or almost one copy of each and every book ever published in modern history![18] And that's just one brain.

If one were to buy a synapse per penny, the entire gross domestic product of the whole world would be barely enough to purchase one human brain. For another comparison, if you could count a neuron per second,

Figure 2.5
Numbers of the brain (for optimal e-book rendering, please adjust the zoom so as to obtain an image as close as possible to uniform gray). Staring closely at this seemingly innocent gray image will reveal a miniaturized checkerboard pattern containing as many as 175,000 black dots (5000 per square inch). The distance between adjacent dots, approximately half a millimeter, is close to the smallest that can be clearly distinguished by naked eye. Even if this entire book contained nothing but gray images like this on each and every page, the totality of those dots would still be insufficient to match the number of neurons in one hemisphere of a single mouse brain. Now imagine as many books as the number of dots in this image, each book with as many pages also as the number of dots in this image, and this same image printed on every page of each of those books. The grand sum of all the dots in this mighty collection would be roughly equivalent to the number of synapses in your brain.

and you did nothing else Monday through Friday 9 am to 5 pm, after an entire 50-year "career" you'd have counted a mere 0.3% of your own neurons. Returning to the botanical metaphor, we should note that the number of neurons in a single human brain is of the same order of magnitude as the number of "real" trees worldwide.[19] Likewise, the number of synapses is comparable to the number of leaves on the planet in a given year.

Although the total number of synapses in the brain is nothing short of flabbergasting, the number of synapses per neuron is very small compared to the total number of neurons. This means that each neuron only contacts but a tiny fraction of all other neurons. Specifically, if every neuron directly communicates with just 50,000 of its 100 billion peers, this sums up to one link every 2 million other neurons, which is very sparse connectivity on a relative scale.

To appreciate what it means for a neuron to contact 50,000 other neurons, consider this exercise. Simply count every person you can think of for whom you could associate a name with a face. To keep things practical, let's say everyone you could identify by name or role (first name or last name, or nickname..., or professional appellative such as "officer," "doctor") and could recognize on the street. You'd be surprised to find out that for the vast majority of people, the resulting number is fewer than one thousand. Most people do not even shake hands with 50,000 other human beings in their entire lifetime! At any one time, a human being is on average "in contact with" one of every 10 million inhabitants on our planet. Thus, not only is the number of nerve cells in our brain over ten times larger than the worldwide human population, each nerve cell is also one hundred times more connected. This roughly yields 1000-fold greater communication traffic in our brain network than in the entire social fabric of twenty-first-century humanity.

One of the most important consequences of the tree-like shape of neurons is that they *invade* a much larger volume than the space they actual *occupy*. For example, it would take a "box" 2 millimeters high, 1 millimeter wide, and half a millimeter deep to enclose the dendritic arborization of a typical neuron in our cortex, but those dendrites would only "fill" less than 0.01% of the box! This means that the dendritic trees of more than 10,000 other neurons are packed in that same volume.

The situation becomes even more extreme for axons. Axons in the human nervous system can reach across opposite sides of the brain, from the brain to the spinal cord, and from the spinal cord through the limb to the extremities of the body. The total wiring of an axonal tree is typically ten times longer than that of a dendritic arbor, but axonal branches are also substantially thinner than dendrites. As a result, the volume *occupied* by the axonal and dendritic trees of a neuron can be fairly similar. The radial span of axons, however, often exceeds that of dendrites by a factor of ten. This means that the volume *invaded* (as oppose to occupied) by the axons of a neuron can easily be a thousand times larger than that invaded by the dendrites of the same neuron. Therefore the "filling fraction" of an axon is one thousand times smaller than that of dendrites, and an axon can share the space it traverses with at least 10 million other neurons. This is a very conservative estimate, as the largest axonal trees can reach out to more than half of the entire brain!

The above numerical exercise might on the surface appear as dry math, but it isn't. In order to form a connection, the axon of a neuron must share the same physical space with the dendrite of its target. Thus, the ability of trees to invade a volume much larger than the space they fill endows them with a tremendous degree of freedom in the choice of their partners. In chapter 6 we argue that this very *potential* to form synapses (powerfully large, but finite nonetheless) defines what we can and cannot learn.

To be more precise, it is not the generic sharing of the same overall space that allows axons and dendrites to make contact but their passing in very close proximity of each other, like the branches of two trees that nearly touch each other. If a neuron makes approximately 50,000 synapses with

Figure 2.6
A didactic wander. (Top) It is easier to explain the continuous integration of information by dendritic branches while walking in a park than in a classroom. (Bottom) The overlapping branches of two trees nearly touching each other offer a useful visual aid to realize that establishing contacts between axons and dendrites requires their passing in very close proximity of each other (photographs by Gerald Goldin).

other neurons, its number of close encounters is approximately half a million (ten times greater). To maintain the metaphor of human interactions within the social network, we only "know" fewer than a thousand people at any one time, but of course we meet new individuals from time to time (and we forget old acquaintances as well). On a typical day we could be introduced to a new colleague at work or meet a stranger at the bus stop. A random encounter is highly unlikely to be someone in a different continent, profession, and social circle. The pool of individuals whom we *could* potentially meet any one day is typically much larger than the number of people we know already but still much smaller than the entire human population.

A popular theme in science fiction involves characters shrunk by a twist of fate and left wandering inside a human body, from the digestive system through blood vessels to lungs and beyond. Many biologists would love to experience such adventures. I feel lucky: in order to pretend to explore my favorite organ from the inside I only need a trail in the woods and a bit of imagination rather than elaborate miniaturizing gadgets. Last Thanksgiving my father-in-law and I took a walk to burn a small fraction of the excess calories we had consumed. Meandering aimlessly in a nearby park, we ended up talking once again about computation in neural networks, a recurrent conversational topic of ours. A scientist himself, he is quite familiar with the abstract neural networks of mathematicians, physicists, computer scientists, and engineers. When speaking with him, I often try to express what the branching structure of neurons implies for the functioning of the *real* neural networks in our brains. During that walk I could point at real trees while hand-waving my description of signal processing in dendrites. I knew this was effective because every few steps my father-in-law would pause, scratch his beard, pull out his camera, and snap a picture of the surrounding vegetation (figure 2.6).

Occasionally during class my students use their smart phones to take a photo of the whiteboard in order to capture a useful explanation. Parks and forests are the best whiteboards to understand the relationship between the brain and the mind. So, if you are so inclined, I recommend you head out to the woods for an inspiring trek. As you grind your way through the next chapters, frequent hikes in the green will reinforce your connections with trees both around and within.

3 Transmitting and Processing Information

3.1 The Axon: Signal Transmission

So far we have painted a picture of the brain as an incredibly busy web of connections, with 100 billion neurons turning on and off, and chemical neurotransmitters carrying messages through thousands of trillions of synapses. Somehow, this bustling action gives rise to, corresponds to, or simply *is* our mental activity. But what exactly constitutes neural information, and how is it exchanged?

As we saw in chapter 1, neurons at rest maintain chemical and electrical gradients across their membranes. These gradients are possible because the neuronal membrane is impermeable to ions. In other words sodium, chloride, and the other cations and anions cannot penetrate the membrane. There are in fact special proteins, *ionic channels*, that form pores through the membranes. These channels are selective for certain ion types. Some channels only let sodium through, others only chloride, and so forth. However, under normal circumstances these channels are closed. Different kinds of channels can be opened by different mechanisms.

The axonal membrane is particularly rich in sodium and potassium channels. These microscopic molecular machines are like tunnels through the membrane with a gate in the middle. The gates are kept shut by the electric tension due to the excess negative charge inside the neuronal membrane.[1] When a neuron becomes active, the electrical gradient across the soma membrane temporarily decreases, allowing the gates to open. Initially, the channels only open up along the part of the axonal membrane attached to the soma,[2] not along the entire axon at once. With the gate in the middle of the tunnel open, traffic begins to move, and the neuron initiates its dynamic process toward communicating with other neurons.

Because sodium is more concentrated outside than inside (refer back to figure 1.2 in the first chapter), it rushes in as soon as the channels open. Sodium is positively charged, so it is also attracted by the excess negative charge inside the neuron. In contrast to sodium, potassium is more concentrated inside. Thus, when the channels open, potassium would tend to move out. However, like sodium, potassium is also positively charged, so its attraction for the negative charges inside initially compensates for the concentration gradient. As a result, the onset of neuronal activity consists of just a rapid sodium entry.

The surge of intracellular sodium decreases the electric gradient across the membrane, eventually leveling out the negative charge inside. This swift change has two important consequences. First, sodium (and potassium) channels in the nearby axonal membrane along the same axon detect the lack of electric tension and they also open up. Sodium then rushes in the adjacent segment of the axon, and the process repeats, triggering a chain reaction that rapidly propagates down the entire axonal tree much like a long series of falling domino pieces.

The second consequence of the electrical gradient leveling brought about by sodium inflow is that there is no excess negative charge keeping the potassium inside. Thus, because of its concentration gradient, potassium starts moving out through its open membrane channels. Because potassium is positively charged, this movement tends to reestablish the original electric gradient. Eventually, when enough electric tension is restored, sodium and potassium channels close their gates.

This cascade of events is called an *action potential* or *spike*. At a given position along the axon, a spike is a rapid sequence of charge reduction across the membrane (depolarization) followed by charge restoration (repolarization), altogether lasting only 1 millisecond (a thousandth of a second). This wave of polarity travels along the axonal tree at a respectable speed of approximately 20 miles per hour. It only takes a tenth of a second for a spike to propagate from the top of the brain to the bottom of the spinal cord (quick, wiggle your big toe, now!).

This mechanism of discharge enables the speedy propagation of axonal action potentials, but it comes at a price. Although after the spike the electric tension is restored across the membrane, the sodium and potassium concentration gradients are spoiled, as sodium moved in and potassium moved out of the neuron. To recover from this state of affairs, neurons have special pumps tasked with chaperoning potassium and sodium ions in and out of the membrane, respectively. These pumps operate continuously even when the neuron is not active, to compensate for small charge leakage across the membrane.

Thus, the axon is a bit like an old-style firearm. The hammer is in tension until the shot is fired, but then the gun must be reloaded. While being reloaded, neurons (just like pistols) cannot fire again.[3] It takes neuronal pumps approximately 1 millisecond to recharge the sodium gradient. This puts a physical limit to how rapidly an axon can fire in sequence.

Maintaining the ionic gradient and restoring it after each action potential by means of pumps consumes a hefty amount of energy. As a matter of fact this is one of the most energetically demanding operations in our body and helps explain why our brain metabolic consumption is higher than that of all other organs, even when we are resting or sleeping. This great energy expenditure also constitutes a nontrivial engineering problem, as heat must be dissipated to maintain the tissue temperature at a physiological level. To avoid overloading our brains, only a minority of neurons in the brain are active at any given time. As an upper bound, if neurons fire on average fifty spikes per second and each spike lasts a couple of milliseconds, only approximately 5% of all neurons in the brain are spiking at any one instant of time. The actual number is likely lower because both the firing rate and spike duration are overestimates.[4]

The peculiar mechanism of information transmission has another important implication. Spikes are not a graded signal but rather an all-or-none phenomenon. This is due to the self-reinforcing action of sodium surge, which lowers the negative charge inside, opening more sodium channels and causing greater sodium increase. In other words if the activity reaches the threshold needed to trigger the chain reaction, it will trigger the same spike whether it was "just enough" or twice as strong. Thus, to a first approximation, the state of the axon can be considered binary: at any one time, the neuron is either firing a spike (active) or not (inactive). Furthermore, once the spike starts down the axon, the same signal is received by all synapses reached along that tree without customization.[5]

This minimalist repertoire might at first seem puzzling. The neuron receives tens of thousands of synaptic inputs and can only respond in one of two ways? That sounds like a very limited message to bother broadcasting to tens of thousands of other neurons! Axons, however, find other means to personalize their signals.

Because of the sparse neuronal activity dictated by energy consumption and heat dissipation, neurons fire at a rate significantly lower than the maximum limit imposed by the duration of the action potential and the pumping recharge period. This minimum signal period (approximately 2 milliseconds) can be considered as the "unit of time" for neurons. If we divide a given time span into a discrete number of these units, we can represent the activity of a neuron as a string of 1's (if the neuron is "on") and 0's (if it is "off"). This digital code resembles the representation of information in traditional computers. This is perhaps the closest similarity one can find between brains and computers among a large number of very substantial differences at all levels.[6]

Despite its simple binary basis the digital code of the axon can carry a substantial amount of information. Neurons can fire their spikes at various rates, from less than once per second up to 500 times per second. Simply counting how many spikes are sent in a given period of time reflects different information content. Moreover, even different spiking patterns at the same firing rate could have different meanings. A neuron might fire each of three spikes respectively at the beginning, in the middle, and at the end of a particular time period; alternatively, within that same time period it might fire three spikes in row and then fall silent, or else it might start quiet and then fire its consecutive triple spike in the middle or at the end of the time period. These four patterns of three spikes could represent distinct messages.[7]

Given the relevance of spike timing, it is also important to consider the slight delays due to spike propagation: from the moment a neuron fires a spike pattern, it might take several milliseconds for synapses to be affected (remember that the duration of the spike itself is about 1 millisecond, and so is the minimum interval between spikes). Synapses closer to the start of the axon will experience shorter delays than those far away along the tree (figure 3.1).

What happens next? When the action potential invades each of the axonal varicosities, it causes neurotransmitters to be ejected into the synaptic cleft. As neurotransmitter dock onto the postsynaptic bank, we shift our attention to dendritic information processing.

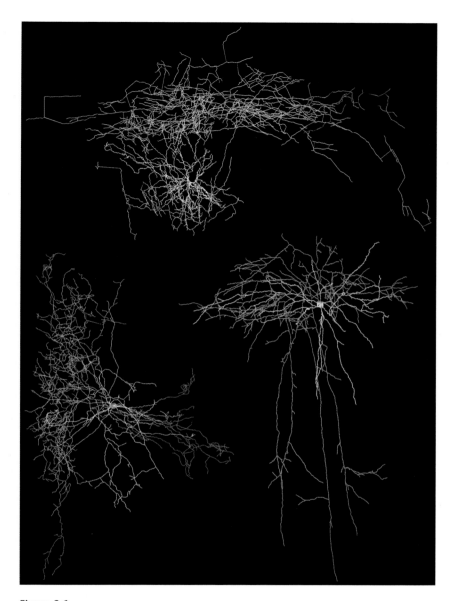

Figure 3.1
Division of labor between axons and dendrites. (Top) A layer 2 spiny stellate neuron from the rat entorhinal cortex.[8] Soma and dendrites are colored white, axon blue (rendering by Amina Zafar in the author's lab). (Bottom left) A CA3 basket cell from the rat hippocampus.[9] Soma and dendrites are colored purple, axon green (rendering by Namra Ansari in the author's lab). (Bottom right) A parvalbumin-expressing interneuron from the mouse neocortex.[10] Soma and dendrites are colored yellow, axon green (rendering by Amina Zafar in the author's lab). These reconstructions are freely available online at NeuroMorpho.org (branch thickness was increased to enhance contrast).

3.2 The Dendrites: Signal Integration

The membranes of dendrites also are rich in ionic channels that normally stay closed. Some of the gating mechanisms for dendrites, however, are different from those for axons. The axonal gates are opened by a pulse of electricity, like an automatic garage door with an electrical switch. The dendritic gates, in contrast, are more akin to traditional key-operated doors. These "keys" are molecules called neurotransmitters. The postsynaptic sites on dendrites have special receptors for neurotransmitters on the outer side of the membrane. Much like keys, neurotransmitters only fit in "their" specific receptors. Whenever the right neurotransmitters bind to the postsynaptic sites, they unlock the channels, allowing ions to pass through the tunnel across the membrane.[11]

The most common neurotransmitters in the brain are called *glutamate* and *GABA*.[12] The glutamate postsynaptic receptors open sodium channels. As you might remember, sodium is more concentrated outside than inside the neuron, and its positive charge is also attracted to the excess negative charge on the inner side of the membrane. For these two reasons, when glutamate binds to its receptors, sodium rushes in through the open channels. This ionic movement creates a temporary reduction in the local electric gradient. Because this change goes in the same direction as the depolarization necessary to start an action potential in the soma and axon, glutamate performs an *excitatory* action on the postsynaptic neuron. For this reason the transient fluctuation in the membrane's electric tension caused by the influx of sodium ions through glutamate-gated channels is called an "excitatory postsynaptic potential" or EPSP ("ee-pee-es-pee") in short. EPSPs are much smaller signals than action potentials. Each EPSP amounts to just a few percent of the electrical jump of a spike.

In contrast, GABA postsynaptic receptors open chloride channels. Chloride is also more concentrated outside than inside the neuron, but it carries a *negative* charge. Thus, when it flows through the open channel, chloride creates a temporary increase in the local electric gradient. This change goes against the depolarization necessary to start an action potential in the soma and axon. Thus GABA performs an *inhibitory* action on the postsynaptic neuron. The transient fluctuation in the membrane's electric tension caused by the influx of chloride ions through GABA-gated channels is therefore called an "inhibitory postsynaptic potential" or IPSP ("i-pee-es-pee") in short.[13] Excitatory synapses are almost exclusively found on dendrites, but neurons can also receive inhibitory synapses directly on the soma or even on the axonal initial segment.

To summarize, EPSPs correspond to the electric gradient *decrease* caused by glutamate, pushing the neuron toward its firing threshold; IPSPs correspond to the electric gradient *increase* caused by GABA, pulling the neuron away from its firing threshold.

EPSPs and IPSPs spread through the dendrites from their local postsynaptic origin. In doing so, these electric signals lose some of their strength and also tend to smear in time and space. Therefore, when they arrive at the soma, signals from distant synapses tend to be weaker and to last longer (up to dozens of milliseconds). Because of the greater distance to cover, synaptic signals from faraway branches will also be received at the soma with greater delays from the time of synaptic activation than those closer to the soma on the dendritic tree.

I like to imagine that each neuron decides whether or not to spike at any one time by vote of all of its synapses, except that the polling mechanism is biased in an unusual way. The votes of every synapse near the soma get counted several times for only one election. The votes of distant synapses get counted for multiple election cycles, but just once at each election, and only starting from the next election cycle. This design makes for quite a strange political system but also apparently some powerful computational machinery!

Dendritic membrane also has ionic channels that are gated by electric tension rather than neurotransmitters. These dendritic channels are similar in gating mechanism to those found in the axons, but they are less abundant and more varied on dendrites. Most voltage-gated dendritic channels are opened by a decrease in the membrane electric tension. They thus tend to be opened by a convergence of EPSPs. Some of these channels are permeable to sodium, which tends to enhance the effects of EPSPs. Others are permeable to potassium, which tends to reduce the effects of EPSPs. A variety of dendritic channels are permeable to calcium, which affects the membrane electric tension similarly to sodium but additionally triggers a series of biochemical reactions inside the neuron. Yet other channels, permeable to different degrees to both sodium and potassium, are opened by an increase, rather than decrease, of the electric tension. These channels are therefore responsive to IPSPs (usually reducing their strength) rather than to EPSPs. This diverse compendium of voltage-gated channels can partially or even fully compensate (or sometimes exacerbate) the loss and spread of synaptic signals as they propagate along the dendritic branches (figure 3.2).

Because voltage-gated channels are less abundant in dendrites than in axons, they do not generate action potentials as strong. However, if enough are activated at the same time, even dendritic channels can generate weak spikes. These dendritic spikes can propagate through the dendrites. Although to a lesser extent than EPSPs, dendritic spikes can also lose strength during propagation, and on occasion they can fail to cross branch points. Even if and when it does reach the soma, a single dendritic spike may not be strong enough to trigger an axonal action potential by itself. Interestingly, when the neuron fires an action potential down the axon, the somatic depolarization may also open voltage-gated channels in the nearby dendritic trunks. As a result, dendritic spikes can also propagate backward into the dendrite, sometimes all the way to the distal branches.

The resulting picture of a dendritic arborization is more akin to that of a hectic street market than to that of a neatly lit Christmas tree. At any one time, thousands of synapses are generating a mix of EPSPs and IPSPs scattered throughout the branches. As all these signals propagate along, they open and close voltage-gated ionic channels on the way, altering their strength and duration as a consequence. Local dendritic spikes are occasionally generated anywhere in the tree as well in the trunk every time the neuron fires an action potential. Each of these spikes also propagates both toward the soma and toward the tips. Somehow, the soma must continuously integrate this whole information to decide at every instant whether or not to fire a spike down the axon.

Despite its chaotic appearance, such bustling dendritic activity constitutes a surprisingly sophisticated implementation of fine computational processes (figure 3.3). As a matter of fact, dendrites of single neurons are capable of performing a broad range of nontrivial arithmetic operations.[16] To a very first, crude approximation, dendrites can simply sum EPSPs and subtract IPSPs. If the net balance of excitatory minus inhibitory changes in electric tension provides enough depolarization to open a critical mass of sodium channels in the axonal hillock, an action potential is generated, otherwise not.

The story, however, becomes more interesting. First of all, not all synaptic inputs are the same. As explained above, signals contributed by EPSPs

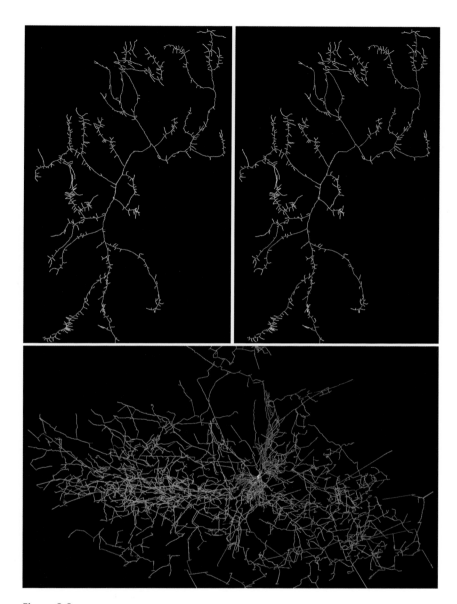

Figure 3.2
Dendritic signal integration. (Top) A sensory class 3 DA neuron from the fruit fly larva[14] colored (left) with the main dendritic skeleton (trunk and branches) in purple and the terminal tips in light blue. On the right image the colors are changed to illustrate the effect of synaptic inputs from the dendritic tips on the membrane voltage of the entire tree (rendering by Namra Ansari in the author's lab). (Bottom) A CA3 pyramidal neuron from the rat hippocampus[15] with dendrites colored red and axon green as a reminder that signals integrated in the dendritic trees are then propagated through the axonal arbor (rendering by Amina Zafar in the author's lab). These reconstructions are freely available online at NeuroMorpho.org (branch thickness was increased to enhance contrast).

Figure 3.3
Tree resembling neuron encoding tree (Daniel Segrè, 2013, Brookline, Massachusetts).

from faraway synapses are usually weaker but longer-lasting. Moreover, two EPSPs on the same branch might result in an integrated signal that is weaker than their sum. The reason is that each of the two also reduces the electric tension in the nearby postsynaptic region, thus reducing the incentive for sodium to rush in from the open channels. At the same time, under appropriate circumstances multiple colocalized EPSPs can also yield a resultant that is *stronger* than the sum of the parts if they jointly manage to open voltage-gated sodium or calcium channels and even initiate a dendritic spike.

Some dendritic signals (EPSPs or local spikes) coming into a bifurcation from either branch are too weak to pass the merge point and would fail to convey their message to the soma. If *both* branches carry a weak signal, however, and the two signals arrive at the merge point at the same time, they may together succeed and proceed as a single combined signal in the journey toward the soma. The specific shape of the branch points on the dendritic tree can affect the probability of transmission. Much like the overall arbors, dendritic bifurcations come in diverse shapes (figure 3.4) and can even change their geometry from time to time.

Furthermore, two spikes advancing in opposite direction along the same branch might interfere with each other when colliding head-on. In the same vein a potent backpropagating spike could pretty much annihilate all incoming signals, weak and strong alike, essentially resetting the entire tree. All in all, the dendritic tree constitutes a powerful processing device. Its main job is to integrate thousands of incoming synaptic signals in parallel and in real time by performing complex computational operations. As a result, dendrites provide the soma with a highly elaborated "recommendation" as to whether and when to fire an action potential.

3.3 The Adaptive Neuron: Plasticity of the Input/Output Relation

Chapter 1 introduced neurons as binary devices that could turn on or off at any time. The challenge, we said, was to figure out how the distributed flickering of a huge number of such neurons can somehow orchestrate our mental life. By now you might start considering the dendritic tree as a miniature computer in its own merit. The individual computing units of the brain might not be neurons after all but rather branches and synapses. Nevertheless, the story so far leaves open a big question: how can our brain

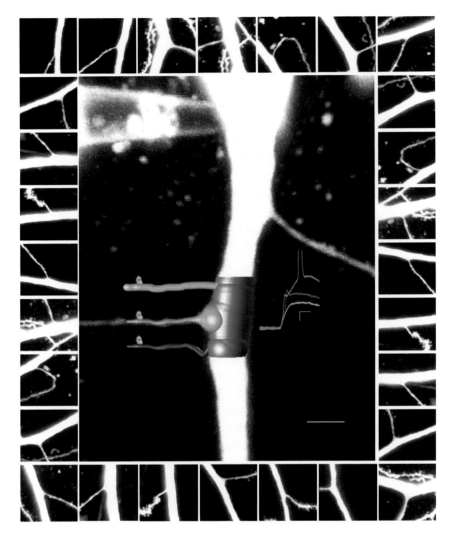

Figure 3.4

Dendritic branching. (Frame images) Natural variability of branch point morphologies in neuronal dendrites from the rat hippocampus, imaged by fluorescent microscopy. (Main image) Graphic illustration of synaptic excitation in three dendritic branches (colored green, orange, and blue) with corresponding voltage signals measured at the soma.[17] The green branch yields a typical subthreshold EPSP, the blue branch propagates a dendritic spike, and the orange branch triggers a full somatic spike (figure by Michele Ferrante in the author's lab). The big dagger-looking spear impaling the neuron from the left at the top of the main image is the microelectrode used for electric recording. The horizontal line on the bottom right of the main image is a 4-micrometer scale bar. The vertical and horizontal scale bars under the voltage traces mark 5 millivolts and 3 milliseconds, respectively.

change its future computations based on the information it continuously processes? *Plasticity* is the term used by neuroscientists to refer to changes in the structure, activity, and function of the nervous system as a consequence of experience.

As it turns out, even the "elementary" components of neurons are not so elementary but possess amazing capabilities for tuning their function depending on complex dynamic interactions. Branches and synapses continuously adjust their responses to incoming information. Some of these changes are short-lived. For example, when a synapse is activated twice consecutively, the response to the second stimulus may be weaker than that to the first. This so-called "pair-pulse depression" may be due to the temporary depletion of neurotransmitters in the axonal varicosity or to the transient saturation of postsynaptic receptors. Other synaptic changes following activation can last from hours to years and are thus referred to as long-term.

In the best known formulation of synaptic plasticity, the coactivation of two nearby neurons strengthens their connection. In other words, if a presynaptic neuron systematically contributes an EPSP to the firing of a postsynaptic neuron, the subsequent EPSPs of that same synapse will be stronger. This is one of the two famous postulates proposed by Donald Hebb (1904–1985); we discuss Hebb's other postulate (about cell assemblies) in the next chapter. In fact, it is worth quoting Hebb's original words[18]: "When an axon of cell A is near enough to excite B and repeatedly or persistently takes part in firing it, some growth process or metabolic change takes place in one or both cells such that A's efficiency, as one of the cells firing B, is increased."

The reference to growth processes may be interpreted as to imply formation of new synapses (a possibility we consider in chapter 5). Most importantly the seemingly innocuous condition that the two neurons must be "near enough" is too often omitted from consideration. We instead elaborate substantially on the cognitive consequences of this essential physical constraint in chapter 6. For now, however, we can comfortably assert the following: even the "simple" interpretation that synaptic strength is increased by the systematic contribution of the presynaptic neuron to the firing of the postsynaptic neuron has long been recognized as a fundamental learning process.

The results of the famous experiments conducted by Pavlov on his dog in the 1920s, for example, may be explained by the process of synaptic plasticity. By ringing a bell every time he fed his dog, Pavlov showed that, after a certain learning period, the dog would salivate just on hearing the bell sound. A second dog, fed on the same schedule as the first, was also presented the same number of bell rings, but at random times rather than paired with food. After those "training" sessions, this "control" animal did not salivate when the bell was rung, although both dogs would salivate equally when the food was brought in. The long-term associations in the first dog's brain may be attributable to an increase in synaptic strength brought about by repeated pairing of corresponding stimuli.[19]

How could Hebbian synapses provide a mechanism for associative memory? Suppose that one neuron in the dog's brain represents food, another neuron represents the bell sound, and a third neuron controls salivation.[20] Before the start of the experiment the food neuron is strongly connected to the salivation neuron, so much so that every time the first neuron spikes (because the dog was presented with food), the resulting EPSP in the salivation neuron is sufficient to elicit an action potential postsynaptically (making the dog salivate). The bell neuron also synapses onto the salivation neuron, but their connection is weaker in the sense that if the presynaptic neuron fires (in response to the bell), the resulting EPSP does not by itself trigger a salivation spike.

Let's now reconsider what happens inside the brain of Pavlov's dog. Because the bell rings when the food comes in, the food neuron and the bell neuron are both "on." But the food neuron activates the salivation neuron. As a consequence, both the (presynaptic) bell neuron and the (postsynaptic) salivation neuron are active at the same time. Because of Hebb's plasticity rule, their synapse becomes stronger, meaning that subsequent EPSPs will be a bit closer to cause an action potential in the absence of other stimuli. After a few such associations, the EPSP goes over the threshold, and a spike in the bell neuron causes a spike in the salivation neuron. From an external observer perspective, the dog learned the association between the food and the bell.

There are many complex and not fully understood mechanisms underlying associative synaptic plasticity. In the next chapter we argue that this kind of synaptic plasticity is more suitable to underlie different kinds of cognitive phenomena than the acquisition of new knowledge. At the same time we propose alternative neural correlates that better explain associative learning and memory. Nevertheless, there is no doubt that synaptic plasticity constitutes a fundamental functional component of brain machinery.

Earlier we introduced two kinds of dendritic ionic channels: those gated by electric tension and those gated by neurotransmission. They are both normally closed, but the former open when the excess negative charge inside the membrane is temporarily reduced, whereas the latter open when the neurotransmitter molecule binds outside the membrane. There is also a hybrid kind of ionic channel in the dendrite, which is gated by *both* electric tension and glutamate. This hybrid channel is also normally closed, but, when open, it is generously permeable to calcium (as well as sodium). Entry of calcium into the postsynaptic region is a powerful event because it initiates a cascade of molecular processes that result in larger subsequent EPSPs. For example, calcium causes chemical modifications of the "normal" glutamate-gated channels, increasing their sodium permeability. Nearby voltage-gated sodium channels may be similarly modified on calcium entry, also becoming more permeable to sodium. Greater sodium inflow means stronger excitation (figure 3.5).

Although powerful, hybrid channels are usually closed even when glutamate is present in the synaptic cleft (which opens other excitatory neurotransmitter-gated channels) or when the neuron is depolarized (which opens other voltage-gated ionic channels). They are only activated by the co-occurrence of two events: the release of glutamate *and* a simultaneous decrease in electric tension. This coincidence corresponds to the synchronous activation of the pre- and postsynaptic neurons, exactly the prerequisite for Hebbian plasticity. Note that the mechanism is highly specific for individual synapses, not entire neurons. Coactivation of two neurons only strengthens "their" joint synapse(s), not the other contacts made by the presynaptic axon onto other neurons or the other contacts received onto the postsynaptic dendrite by other axons. It is not even necessary for the postsynaptic neuron to fire an action potential: a local branch depolarization, caused for example by a dendritic spike, even if it fails to propagate to the soma, is sufficient.

There are many other forms of long-term synaptic plasticity. Sometimes a rapid sequence of action potentials (referred to by neuroscientists as a "burst") fired by an axon can also strengthen a synapse. Suppose for example that a neuron fires a burst of twenty spikes one right after the other.

Figure 3.5
Active dendrites (photograph by Daniel Segrè).

The first few spikes (say, spikes one to ten) depolarize the postsynaptic dendrite. While the dendrite is still depolarized, the next ten spikes arrive (the other half of the burst). The coincidence between the depolarization (due to the first ten spikes) and the immediately subsequent presynaptic firing (the end of the spiking sequence) provides the condition for synaptic reinforcement.

Mechanisms also exist to decrease synaptic strength (long-term depression). Specifically, postsynaptic depolarization just *before* activation of the presynaptic neuron (rather than just after) renders the subsequent EPSPs weaker. The exquisite sensitivity of the *direction* of synaptic plasticity to the temporal sequence of events highlights the important information content of exact spike timing in axons.

Another form of plasticity is mediated directly by particular types of neurons that work as "neuromodulators." These neurons alter the way their postsynaptic targets respond to the excitatory or inhibitory inputs from *other* neurons. These mechanisms underlie circadian rhythms such as the sleep-wake cycle and also more transient states of alertness or mood. Modulatory neurotransmitters bind to special kinds of receptors that do not directly open ionic channels. Instead, when bound to a neurotransmitter, these receptors trigger chemical reactions similar to those initiated by calcium entry. Such molecular events can also alter the production of specific proteins (the constituents of ionic channels, neurotransmitter receptors, and much of the rest of the neuronal machinery) and even affect genetic expression in neurons. These consequences are very long-term and can last for the entire life of an individual, just as can the view from the window in the first-grade classroom or one's inclination toward mystery novels.

3.4 The Mighty Cable of the Axon

Dendrites might already appear as the computational powerhouse of neurons. The richness of their vocabulary, from small excitatory and inhibitory signals to local spikes and from neurotransmission to biochemical cascades, endows dendrites with the ability to process the massively distributed synaptic chatter they receive with orderly mathematical operations. In contrast, the axon is tasked with the seemingly trivial responsibility to transmit this logically integrated information to other neurons in the form of simple zeroes and ones by firing or not. Our brief survey of the adaptive capability of neurons may have exaggerated this contrast. After all, much of the action related to synaptic plasticity occurs in dendrites.

Although from the point of view of an individual neuron the above description seems fair, a different perception emerges from the perspective of the entire network. A prominent feature of axons is their sheer length (figure 3.6). In some "local" neurons the axonal tree can be just about the span of the dendritic arborization, yet the axon tends to be much "bushier" (figure 3.7). The contrast becomes even more extreme in "principal cells," which (as their name suggests) constitute the majority of neurons. The axons of principal cells have an overwhelmingly greater reach than their dendrites. If we enlarged one of these neurons so its dendritic tree is roughly your height, its axon could be higher than the tallest skyscraper in the world. If we enlarge the dendrite to the height of an alpine tree to reinstate our useful botanical metaphor, the axon of the same neuron would extend from the timberline[22] all the way down to the base of the mountain at sea level. For a more suburban visual metaphor, imagine the dendritic arbor is a backyard tree extending its branches over the roof of the house. The axon attached to the same soma would extend its branches over the roofs of houses more than twenty-five blocks away.

Even more telling than the height or branch span of an axonal arbor is its summed cable length. Axonal branches constitute more than 95% of all the wiring in the brain,[23] including dendritic trees, glial cells, and blood vasculature (arteries, veins, and capillaries). To imagine how long all the branches in the axon of a single neuron are if summed together, there is no need to mentally enlarge the neuron to the size of a botanical tree. Even leaving the measures to the actual 1:1 scale, the overall branching length of a typical axon sums up to a yard or more! These microscopic structures wind along for macroscopic distances that far exceed the width of the human head from ear to ear.

To appreciate how stunning this measure is, consider the total length of all axons in a human brain: about 50 million miles.[24] This is more than twice the length of all paved roads of the entire world![25] If you don't find that impressive enough, try this: if you were to unfold all the axonal wiring of a single human brain and wrapped it around the equator, you would circle the Earth more than 2000 times before running out; it would take ten years for a Boeing 747 to fly that distance continuously, without accounting for take-off, landing, or refueling. To put it another way, the axons of a

Transmitting and Processing Information

Figure 3.6
The mighty cable of the axon (Daniel Segrè, 2014, Fresh Pond Park, Cambridge, Massachusetts).

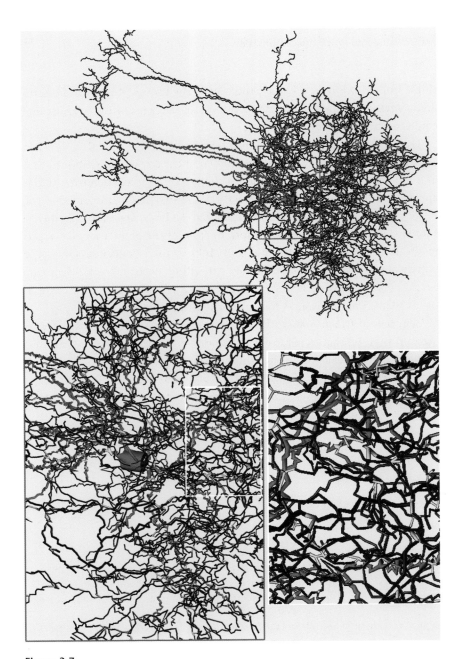

Figure 3.7
So tiny, yet so long. A neurogliaform interneuron from the rat motor cortex.[21] The soma and dendrites are colored magenta; the axon is in shades of blue with yellow bifurcations. The left and right bottom insets display progressively more zoomed-in details of the arborization (rendering by Amina Zafar in the author's lab). The span of this neuron is less than half of a millimeter in height and a third of a millimeter in width, but the total axonal length is over 40 millimeters (more than 1.5 inches). The reconstruction is freely available online at NeuroMorpho.org (branch thickness was increased to enhance contrast).

single human brain could go all the way to the moon, and back one hundred times. Stitching together the axons of an average married couple, one could reach all the way to the Sun.

These numbers are particularly noteworthy because, again, they apply to axonal cables as they are, not scaled to match the size of their arbor to that of a botanical tree. Recall that the real height of a typical dendritic tree is approximately 1½ millimeters or the thickness of a penny. Measured as the height of piled-up pennies, the total cabling of a brain (again mostly due to axons) is worth some US$500 billion. In spite of the greater length, axons in the human brains are thinner, not thicker, than dendrites. Using the botanical metaphor, those branches extending twenty-five blocks away would be as thin as shoe strings. The brain can pack up so much cable in the available volume thanks to the axons' ultraslim design.[26]

In terms of functional worth the impressive length of axons redeems them in their imaginary rivalry with dendrites. After all, the most quintessential feature that distinguishes the nervous system from all other organs is network connectivity. The fabric of this web is woven out of the neuronal branching cable. It is thus axons, not dendrites, that bear the lion's share of the responsibility for keeping the brain interconnected.

Axons and dendrites are thus distinguished by both sharp functional and structural differences (table 3.1). The role of the spectacularly long

Table 3.1
Properties of axons and dendrites

Property	Dendrites	Axons
Signal direction	Neuronal input	Neuronal output
Number of stems	Several trees per soma	Only one tree per soma
Relation to synapse	Postsynaptic (receiver)	Presynaptic (sender)
Mode of signaling	Mixed passive/active and graded	Fully active and all-or-none
Typical total length	A few millimeters per neuron	Several centimeters per neuron
Branch geometry	Strong taper, many bifurcations	Weak or no taper, many bifurcations
Myelination	Never wrapped in myelin	Myelinated in white matter
Fine structure	May be covered with spines	Typically dense in varicosities
Synaptic ultrastructure	Contain neurotransmitter receptors	Release vesicles with neurotransmitters

axons is to convey simple messages reliably and quickly over long distances to tens of thousands of specific targets throughout the brain. Such information transmission is mediated by a digital code of all-or-none spikes. In contrast, dendrites are powerful computing devices that perform complex operations by integrating tens of thousands of graded signals from other neurons. Because the axons of these other neurons cover most of the brain distance, dendrites can afford to carry out their information processing within a spatially contained tree. This is fortunate because signals are continuously changed as they advance through the dendritic branches and would likely be lost to noise over longer distances.

The distinct architectural arrangements of axonal and dendritic trees reveal a striking correspondence between evolved structural features and optimally fitting functional roles. An interesting "thought experiment" may be useful to appreciate fully the structure-function relationship in a direct comparison between axonal and dendritic arbors. Suppose for the sake of argument we build a brain network by maintaining the very same physical organization of neurons, but we invert the direction of information flow. The resulting long trees (previously axons) would function as input cables (let's call these hypothetical mega-antenna receivers "dendrons"). The stubby arbors (previously dendrites) would serve as spatially contained loudspeakers (which we will name "axites").

In order to "hear" signals from far away all the way at the soma, dendrons would have no choice but to adopt the same biophysical machinery of active conductances that we discussed for "real" axons, leading to all-or-none spikes. Unfortunately, this design would not allow neurons to discriminate among the rich combinations of possible synaptic inputs from the tens of thousands of presynaptic neurons contacting the dendron. With the high density of voltage-gated sodium (and potassium) channels necessary to transmit information over such long distances, the compounded synaptic signal would either exceed the spiking threshold or not. In the former case it would produce an action potential independent of whether it had barely reached the threshold or passed it twice over. In the latter case the signal would die off before reaching the soma, independent of whether it was just shy of the threshold or it was a flat line.

Axites, in contrast, would not need so much active conductance to transmit their messages to the postsynaptic neurons. However, without systematic all-or-none spiking, the output of axites would lose consistency and reliability. Different targets would receive signals of variable strengths. The contacts made by axites near their soma would sound louder than those made farther in the tree. Obviously, this wouldn't work quite as well as real brains.

One could even conceive of an "intermediate" design," with input and output trees of similar length—and even some neurons with long input *and* output trees and others with short input-output trees. But the same line of reasoning followed above easily shows that those alternative designs wouldn't work either. Both reliable output signaling and long-distance communication share the same requirements of active all-or-none spiking or digital information transmission. Likewise, powerful input integration demands flexible signaling mediated by analogic (continuous) processing, which puts a hard limit on the spatial extent of the underlying cable. This also explains why, in complex brains, inputs and output of neurons are segregated in physically separate trees, as opposed to being shared in one and the same bidirectional cable. As the first section of chapter 7 shows, the structural and functional separation of axons and dendrites is not necessary in simple nervous systems such as those of worms.

Part II: Dynamics of the Brain-Mind Relationship

4 Activity Patterns and Mental States

4.1 The First Principle of the Brain-Mind Relationship

The previous two chapters overview the main actors, mechanisms, and principles of the brain at the scale of its cellular constituents, the neurons. It is now time to tackle the hard question: How does the brain relate to the mind? What links the structure and activity of neuronal networks to thoughts, feelings, intentions, memories, perceptions, and dreams? A definitive solution to this problem is not only missing but not even close. In fact, the brain-mind relationship is generally viewed as the ultimate, perhaps greatest scientific challenge for humanity. Here we are not just referring to the behavioral outcomes of brain activity, such as movement and speech. We are including the intimate contents of unspoken thoughts, untold dreams, and never executed intentions.

Although theories exist linking brain structure and activity with mental function, they are either grossly incomplete in terms of what they can test and predict or they have not yet gained vetting of scientific validation and acceptance of the research community. For the vast majority of current theories, both of the above hold true. Nevertheless, there appears to be a relatively broad consensus on at least a basic principle underlying the brain-mind relationship. This principle is seldom stated explicitly, but most scientists would probably agree that *mental states correspond to patterns of electric activity in the nervous system*. It is difficult to imagine a plausible alternative to this general formulation.

To make things a tad more tangible we step up the claim by specifying that the electric activity is expressed in terms of neuronal spikes. This is perhaps the most likely and obvious choice for the neural correlates of mental states because spikes constitute the main carrier of information exchanged among neurons. Moreover, by signaling the on/off state of neurons, spikes capture the spatial and temporal scale most relevant to cognitive dynamics. Thus, I believe that the majority of active neuroscientists, if asked, would agree with the selection of neuronal spikes as the essential signals underlying consciousness. At the same time, there are other possibilities that cannot be logically or empirically discounted at present. For example, subthreshold oscillation, synaptic currents, and local dendritic potential are also constituents of "electric activity in the nervous system" and cannot be ruled out as neural correlates of mental states. However, as a working hypothesis, we state the following first principle of the brain-mind relationship:

Mental states correspond to patterns of spikes in the nervous system.

We postulate this principle to apply to any and all mental states (see section 1.4). Thus, for a given brain, different spiking patterns correspond to different mental states, and vice versa. The spike "patterns" are taken to possess both spatial and temporal extent, meaning that they are determined by both *which* neurons are firing and *when* they are firing. We already know that distinct aspects of our mental content are encoded by specific regions of the brain. For example, the sight of an approaching car and its sound correspond to spiking patterns involving neurons in different brain locations. The sight of *another* car would be encoded in the same part of the brain as that of the first car but either by different neurons in that region or by a different temporal sequence of spikes in the same neurons or, most likely, by both nonidentical sets of neurons and spiking times.

To ground the notion of spiking pattern with a familiar metaphor, it is useful to connect this concept to the music produced by a full symphonic orchestra.[1] There are many instruments of different kinds, some more different than others (say, winds, percussions, and strings), and also multiple instruments of the same kind (such as violins). Here each instrument may be compared to a neuron (recall that neurons come in different types, shapes, and sizes, as do botanical trees). At any one moment each instrument is playing one note or accord, but the symphony is truly a collective sequence of such instants: the melody would be unintelligible from one brief point in time or if only the bass were playing. In an orchestra the same music can be played to a certain extent faster or slower, but beyond certain limits it would no longer be recognizable. Similarly, a spiking pattern can be

Activity Patterns and Mental States

imagined "played" faster or slower (perhaps corresponding to different speeds of the experienced mental state) but only up to a certain point before losing its meaning.

Many mental states are complex constructions of simpler mental states. In the previous example the perception of an approaching car has both a visual and an auditory component. It is possible to separate the two, for instance, by closing your eyes to isolate the sound or plugging your ears to isolate the sight. Because the two sensory components co-occur in the full percept, it is reasonable to presume that the corresponding spiking pattern is a direct composition of those encoding the separate mental aspects. Similarly, most mental states are formed by sequences of shorter mental states. The approaching car, for example, gives rise to a different feeling when it is still far away than when it is getting very close. Again, it only makes sense to imagine the spiking patterns of an extended mental state as the sequence of each of the shorter patterns that underlies the corresponding mental state.

It is interesting to ask if there is a "minimum" spatial and temporal extent of spiking patterns. Logically, it is possible that a single spike in an individual neuron could underlie a distinct mental state. This is unlikely, however, to be the case. The simple notion that each neuron would represent a different idea is known as the "grandmother cell" hypothesis, indicating a putative neuron in everyone's brain firing selectively every time granny is thought of. Interestingly, these kinds of neurons have indeed been identified in the human brain.[2]

However, this doesn't mean that these cells act alone. In a musical metaphor, *cannons* may well be the identifying "instruments" of Tchaikovsky's 1812 Overture, in that there is no other famous piece of symphonic score employing them for sound effects. Nevertheless, the booming of cannons would hardly be sensible without the rest of the orchestra playing a musical score. The same appears to hold true for grandmother-like cells; in each and every case that was ever tested, information was always found to be represented by groups of neurons instead of single cells.

In all mammals from mice to humans, for example, the region of the brain called the hippocampus contains neurons that signal specific locations in the environment, a bit like GPS devices. These neurons are aptly referred to as "place cells" (figure 4.1). Some hippocampal place cells in your brain, for instance, might fire selectively when you are standing

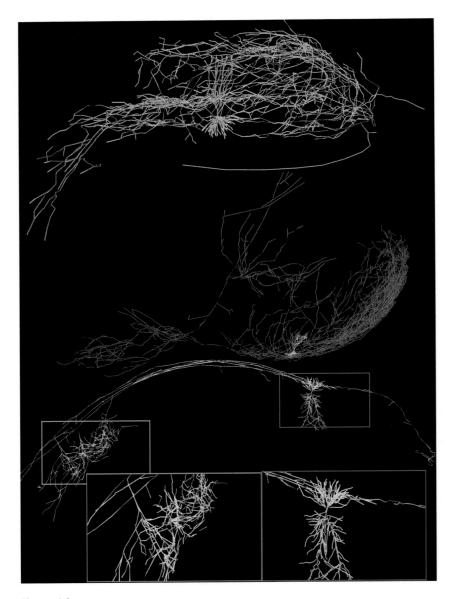

Figure 4.1

Place cells: neurons encoding space. (Top) A CA3 pyramidal cell from the rat hippocampus. The basal and apical dendrites[3] are colored blue and green, respectively, with yellow terminal tips; the axon[4] is colored brown (rendering by Uzma Javed in the author's lab). (Middle) A CA2 pyramidal neuron from the rat hippocampus (see note 4). Dendrites are colored white, axon red (rendering by Amina Zafar in the author's lab). (Bottom) A CA1 pyramidal cell from the rat hippocampus. The basal and apical dendrites[5] are colored green and pink, respectively, with dark blue bifurcations; the axon (see note 4) is colored light blue. The two insets zoom in, respectively, on the axonal terminals and the dendritic trees (rendering by Namra Ansari in the author's lab). These reconstructions are freely available online at NeuroMorpho.org (branch thickness was increased to enhance contrast).

in front of the stove in the kitchen, and others while you are in the shower. The key point here is that there are *many* "stove" cells and many "shower" cells in your hippocampus. In other words, more than one neuron responds to each location.[6]

This evidence supports the notion of *distributed representations* in "cell assemblies" first proposed by Donald Hebb. Cell assemblies constitute Hebb's second landmark contribution to modern neuroscience.[7] It is not known how big cell assemblies actually are: the number of neurons in a cell assembly likely varies broadly among brain regions, from a few dozens to several thousands.

Similarly, a single spike time in even a large cell assembly cannot underlie a mental state any more than a few notes played by a full orchestra would sound like a symphony. Recall from section 1.4 that an experience must last at least a certain amount of time (approximately 50 milliseconds) to be consciously accessible. We should therefore presume that the same temporal requirement applies to the spiking patterns underlying conscious experience. Because a spike lasts about 1 millisecond, and it takes the neuron also roughly a millisecond to recharge (as explained in section 3.1), we can safely conclude that the briefest mental states are encoded by up to twenty-five spikes per neuron.[8]

We can thus envision an "elementary" spiking pattern as the activity of a cell assembly (typically of a couple of hundred neurons) for up to twenty-five spikes per neuron (but on average much less) over a 50-millisecond period. If each of 200 neurons fires on average five spikes, we have a choreographed distribution of one thousand spikes lasting only one-twentieth of a second. We propose this cell assembly activity to be the neural correlate of a "moment of awareness."

4.2 We Experience Only a Minute Fraction of Our Possible Mental States

On the surface the first principle of the brain-mind relationship seems rather uncontroversial, perhaps almost trivial. Yet further considerations highlight its deep and long-ranging implications. An important consequence of the first principle is that we experience only a minute fraction of our possible mental states.[9] In other words, the mental states actually experienced by any individual over the course of his or her lifetime amount to just a small sample of all mental states that are *actually* accessible to that person.

Figure 4.2
Everything and the opposite of everything (Daniel Segrè, 2014, Halls Pond Sanctuary, Brookline, Massachusetts).

Activity Patterns and Mental States

The face value of this assertion is fairly obvious: at any one instant we experience only a single mental state, and our lifetime has a finite number of mental instants (see section 1.4). But how many total mental states *could* we experience at any one instant? The first principle provides a useful neural correlate: the mental states we *could* experience (figure 4.2) correspond to the spiking patterns that could be instantiated by the neural network in our brain.

On the one hand, at a given moment only one particular spiking pattern is active in each brain, out of the many more spiking patters that could be active in that brain. On the other, the entire collection of spiking patterns that could be instantiated in that same brain in this moment is itself a small proportion of all combinations of firing sequences that are in principle possible in that very large set of neurons. For example, if two neurons strongly inhibit each other, a pattern in which they both spike may be unlikely or impossible. In contrast, one neuron strongly exciting another will be conducive to patterns in which activity of the former is followed by activity in the latter. Likewise, if two neurons are not connected to each other and they do not share any incoming connection (that is, they are neither excited nor inhibited by the same set of other neurons), it would be highly improbable for them to systematically fire together.

An intuitive explanation can be thought of in terms of traffic patterns and how they depend on the configuration of the roadway system. After watching a busy intersection for a few minutes, imagine "picking" a car and momentarily closing your eyes. When you open them you will still be able to "look" for the car where you expect it based on the street layout: for example, it may have (1) turned right, (2) crossed the intersection and continued straight, or (3) stopped for a red light. Indeed, as you gaze around you would find out that the car did in fact follow one of these three possibilities. Everyone would (rightly) consider it impossible to find the car on a parallel road that was not even connected to the intersection. The path that the car has taken is one of several it could have taken, but the number of paths it *could* have taken (given the existing roads) is itself much smaller than *all* paths that you could imagine if you disregarded the grid of asphalt. If you followed that car from a satellite (without seeing the road grid), you could initially expect any path, and the car would take just one. If you followed all cars for a sufficiently long time, all their paths combined would effectively redraw the street map.

If you are asked to think about your childhood for a second or two, many memories *could* come to mind, but only one will likely emerge in your consciousness. Yet, if you never played violin, it is impossible that one such mental state would be a recollection of your third-year recital. Note that this is not so much so because *you did not have* a third-year recital but because you could not even imagine what it would feel like! If you *had* played violin but still did not have a third-year recital (because your parents took you out of town that day or because you were sick), reminiscing about your childhood might make you daydream of, if not even falsely remember, your (never happened) third-year recital. If you were a violin player, both remembering and daydreaming about that recital are possible experiences, even if in one case a recital indeed occurred (and was likely witnessed and experienced by other individuals) and in the other it is purely the fruit of your imagination.

To summarize this line of reasoning in more general terms, our experiences are a small subset of possible experiences. The realm of possibility can be thus described as follows:

Many experiences are possible

Every individual could have a few of those possible experiences

Any individual actually has a small subset of his/her possible experiences

A few of these correspond to actual events (shared by others)

The above description assumes experiences as the starting point. Because experiences belong to the "world of ideas," this can be deemed an "idealist" view of reality. In the next chapter we contemplate a complementary "taxonomy of possible existence" based on a materialist perspective (that is, starting from physical events that belong to the "world of matter"). For the time being we continue discussing other important consequences of the first principle of the brain-mind relationship.

4.3 Knowledge Is Encoded in the Synaptic Connectivity of the Network

Why is it, then, that I can *imagine* my third-year violin recital that has never actually happened only if I *am* a violin player?[10] More generally, why are the mental states that each of us is capable of experiencing so tightly linked to what we know (if anything) about the related experiences? As discussed above, the firing pattern corresponding to the mental state of a third-year violin recital can be instantiated only if the corresponding neurons are wired in the network so as to allow that particular flow of activity. Returning once again to the traffic metaphor, consider the neural correlate of a mental state as a given set of "driving paths" for a number of cars. The resulting traffic pattern can possibly occur only if the underlying roads are laid out and connected so as to enable it.

An extension of this argument leads to the conclusion that the connectivity of the network provides a necessary set of constraints to determine which spiking patterns are possible and which are not. In other words the set of spatial-temporal patterns of activity that a circuit can instantiate is determined by its wiring diagram. The graph of all synaptic connections of a person's nervous system is referred to as his or her *connectome*. Thus, according to the first principle, the connectome constitutes a map of one's knowledge, meant as the collection of mental states that said individual can experience. In this sense "to know something" means to be able to instantiate the mental state that represents the corresponding concept. This idea reflects the notion that an individual "is" her or his brain connectome.[11]

The above view explains the current emphasis in neuroscience on mapping the connectome of the nervous system, a field that is sometimes referred to as *connectomics*. In fact, the term "connectome" is typically used with two distinct meanings. At one level, researchers seek to map the *connectivity among regions of the brain*. This is, for example, the goal of the Human Connectome Project, a recent initiative supported by the US National Institutes of Health.[12] This "macroscopic" sense of the word connectome is also termed *projectome* because it is based on regional *projection patterns* rather than actual connections.

Within the cerebral cortex alone, for example, there are at least hundreds of distinct functional regions. You may recall from section 1.2 that the cortex can be divided into four main lobes: the occipital cortex (in the back), mostly dealing with vision; the parietal cortex (at the top), encoding for touch or somatosensation; the temporal cortex (on the sides), associated with hearing and representation of objects and faces; and the frontal cortex (located, well, in front), carrying out executive functions such as decision making, planning, and motor control as well as logical reasoning. Moreover, there are a right and a left hemisphere, each of which usually reflects the opposite side of the body or the environment. So, for example, the right occipital lobe represents the left part of the visual field and vice versa, the left parietal lobe activates when something touches the right part of the body (and vice versa), and each frontal lobe controls the other side of the body. These regions are connected in specific ways; for example, a projection pattern connects the visual cortex with the cortical area representing faces. In broad terms, the projectome aims to capture the overall patterns of connectivity among brain regions.

Each of these regions, however, can be divided further. One major part of the visual cortex, for example, is made out of many smaller subregions, each taking care of a small portion of the visual field. For instance, one subregion encodes for the top left corner of what one can see. Each of these subregions is itself divided into alternating stripes, one representing the information coming from one eye and the next from the other eye. Interspersed within those there are small areas, called *blobs*, that carry the color information of the tiny element of space looked at. Another part of the visual cortex has subregions that encode for edges (useful to outline the shape of objects). Each adjacent little area is active when one is seeing an edge at a particular angle but not at other inclinations.

Similarly, both the motor cortex and somatosensory cortices are divided into small regions each representing different parts of the body, from every finger and toe to patches of skin and muscles on the stomach and back. These regions do not always have sharp boundaries because their activity transitions continuously: for example, the patch of visual cortex encoding 75-degree edges will be maximally active for lines at this orientation, still fairly active at 70 or 80 degrees, somewhat active at 65 or 85 degrees, and completely inactive for edges orthogonal to the "preferred" view.

Because of the lack of objective landmarks for defining brain areas, the regional division is in practice determined by the method of investigation and its instrumental resolution. The study of the human brain is largely limited to noninvasive technologies such as magnetic resonance imaging. Top-of-the-line instruments typically achieve a spatial resolution of 1 cubic millimeter (1 mm^3). Because the volume of the human brain averages 1.5 liters, with current brain scanners it is possible to create a connectivity map among up to roughly 1.5 million small brain regions. We return to human brain scanning in chapter 9.

At some level this connectivity map might be sufficiently fine-grained to characterize mental content, for example, to quantify the brain activity giving rise to all possible movement of our body.[13] However, at a finer level it is expected that projectomes will be insufficient to capture even the mental states corresponding to single words, let alone all the subtleties of mental content (we expand this discussion in section 9.3). In other words, connectivity among brain regions is too coarse to allow associating specific mental states with neural assemblies. Each little 1-mm^3 brain region contains more than 50,000 neurons on average (100 billion neurons divided by 1.5 million regions).

If firing patterns ultimately identify mental states, it stands to reason that the greatest constraints to mental capacity (that is, knowledge) would be provided by the "finer-grained" connectivity map among neurons. This leads to the second meaning of the term connectome: actual connections typically from the axon of one neuron to a dendrite of another neuron, for each and every neuron in a brain (figure 4.3).

This microscopic "connectome proper" is sometime referred to as *synaptome* to stress the foundational synaptic nature of neural connectivity. The most direct technique to identify synapses is electron microscopy. Unfortunately, electron microscopy is a destructive technique, and there is no prospect in sight to deploy it noninvasively in a live, behaving subject. Most importantly, the superior resolution of electron microscopy is offset by a very narrow field of view. Even though electron microscopy can in principle visualize each and every neuron, with all their axonal and dendritic trees, such a feat is typically limited to a volume of less than a single cubic millimeter.

As of now, we are many orders of magnitude shy of the possibility of scanning an entire human brain at the electron microscopy scale. Nevertheless, many recent developments in optical microscopy, and especially confocal fluorescent microscopy, provide the hope that it may relatively soon be possible to image an entire mammalian brain (e.g., that of a mouse) with every axon and dendrite of each neuron. Optical microscopy does not always allow the positive identification of all synapses because its resolution is not as high as that achieved by electron microscopy. However, confocal fluorescence microscopy allows the detection of axonal and dendritic branches. The tight proximity of axons and dendrites could already yield an informative "draft" description of the neuronal connectome. We discuss these prospects toward the end of our journey in chapter 9.

How are the "macroscopic projectome" and the "microscopic synaptome" related to each other? If two brain regions are interconnected, there

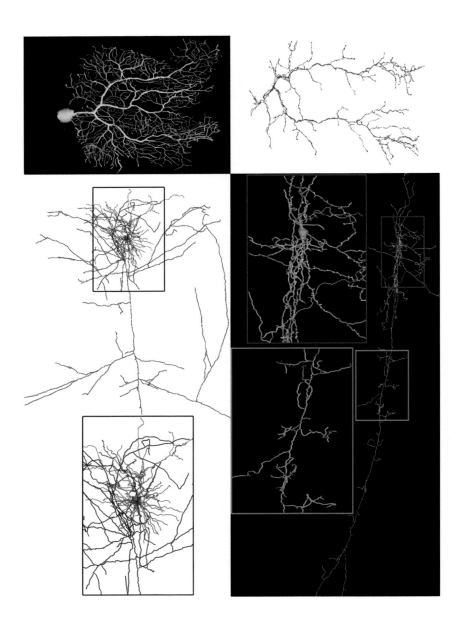

Figure 4.3
Synaptic partners. (Top) Connections in the cerebellum. A rat Purkinje cell[14] (left), with soma and dendrites colored yellow, blue, and green (rendering by Uzma Javed in the author's lab). This neuron type receives connections from climbing fibers (see note 12 in chapter 2) (right) formed by the terminals of axons coming from neurons in the medulla (data by the author). Neither the axons of the Purkinje cells nor the dendrites (and somata) of the medulla neurons were reconstructed. (Bottom) Connections in the neocortex. A layer 2/3 pyramidal neuron from the rat somatosensory cortex (see note 2 in chapter 2) (left). The soma and dendrites are colored black, the (only partially reconstructed) axon red. The inset zooms in on the arbors near the soma (rendering by Amina Zafar in the author's lab). This neuron type forms both pre- and postsynaptic connections with "double-bouquet" interneurons, an example of which is shown on the right (see note 21 in chapter 3). The interneuron's soma and dendrites are colored purple with yellow terminals, axon teal with blue terminals. The blue and orange insets zoom in on the arbors near the soma and on a more distal axonal region, respectively (rendering by Namra Ansari in the author's lab). These reconstructions are freely available online at NeuroMorpho.org (branch thickness was increased to enhance contrast).

must be at least some neurons from one region projecting their axons all the way to the other region. These long-distance cells constitute the majority of the neurons in the most intensively studied parts of the brain. There are also other "local" neurons (often referred to as *interneurons*) that only make outgoing connections to nearby cells, although they can receive incoming signals from projection neurons in other brain regions. We return to this distinction in chapter 8.

Let's just recapitulate the take-home point so far. We propose that synaptic connectivity in the brain is the neural correlate of what a person "knows." The classic perspective on connectomics is that the brain is a network, and therefore, its basic "blueprint" is necessary to understand how it functions. Here we support a stronger claim: that the full map of the brain connectivity at the neuronal level provides a snapshot of the corresponding mind's repertoire. This claim is a consequence of the first principle of the brain-mind relationship because firing patterns are routed through the available wiring.

4.4 A Far-from-Complete Engram

Although physical connectivity poses a "necessary" constraint onto the selection of possible mental states, the biophysical properties of the network (e.g., the sign and strength of synapses) must also be specified in order to determine which firing patterns are possible and which are not. Nevertheless, we can apply much of our reasoning not just to "simple" structural synapses but more broadly to functional connectivity.

Specifically, we can say that a neuron is *functionally connected* to another neuron if the firing of the first neuron systematically increases or decreases the probability that the second will fire immediately afterward. Associating the complete neuron-level functional connectome of a brain with the entire knowledge available to the mind is more rigorous and accurate than doing the same with the physical connectivity map. However, it also complicates this framework both in theory and in practice. Conceptually, it requires switching from "yes or no" binary connections (two neurons are either connected or not) to a continuous influence: a neuron can increase or decrease another neuron's firing probability by a lot or just a little. Empirically, it demands recording electric activity from every neuron for a long time in a functioning brain rather than inspecting its structure. At the same time, among the brain properties that can be practically measured, physical network connectivity provides the most informative indication of the firing patterns that can be instantiated in the circuit because of the very large number of neurons and synapses in the brain. Therefore, structural connectivity is a useful proxy for the neural correlate of retrievable knowledge.

Even with the distinction between synaptic and functional connectivity, however, the picture relating the brain and the mind is clearly still incomplete. First of all, not all characteristics of firing patterns are expected to directly relate to mental states. Some aspects are likely to be in place in order just to keep the network functioning. Other observable features of the firing patterns might even emerge from the mechanisms in place (selected by evolution as fit to do the job) without being relevant to subjective awareness at all![15] By analogy, let's consider an automobile. Some of its parts clearly serve the main purpose of transporting the driver and passengers around: the combustion chamber to burn gas, the pistons to generate movement, the transmission to relay it, the tires to move on the road, the steering wheel and pedals to control speed and direction, and so forth. Other parts of the car, however, simply keep the rest of the engine working, such as the lubrication and cooling systems, or keep the driver and passengers comfortable, such as the air conditioning and radio.

Currently neuroscientists do not know which aspects of firing patterns are important to represent mental states and which are not. A prime illustration of this state of affairs is provided by the wide range of interpretations and hypotheses regarding the phenomenon of rhythmic oscillations in the brain.[16] The activity of neurons, of their assemblies, and of entire regions of the nervous system is known to follow periodic waves of increased or decreased intensity at a variety of frequencies. These oscillations can be detected, for example, by electroencephalography (EEG).

Several researchers think that some of these rhythms (e.g., in the gamma frequency) are the very neural correlate of consciousness. In other words when the firing of a given collection of neurons is gamma-modulated, the subject becomes aware of the content represented by that assembly. Other scientists, however, believe that gamma oscillations are a mechanism to bind together features of the same percept, seamlessly allowing us, for instance, to assign a face, a voice, a smell, a name, and a memory all to one and the same unitary concept of a given person. Yet others in the field view gamma rhythms as nothing more than a side effect of the interplay between excitatory and inhibitory neurons in the normal operation of the network. In practice, scientists don't have consensus on how gamma rhythms, or any other characteristics of the firing patterns, may correspond to mental state.

Another major gap in our understanding of the relationship between firing patterns and mental states is that we are as of yet almost entirely clueless as to what pattern maps onto which mental content. Another way to express the same issue is that we are missing the code to translate firing patterns into mental states (and vice versa). We know the coarse correspondence between brain regions and overall "theme" of the representation. For example, we can identify a cortical area for language production, one for language comprehension, one for recognition of faces, and so on. But we are not even close to having the spatial and temporal resolution to discern which neurons are firing within those areas and how these patterns code for specific words, sentences, and faces.

An even more foundational issue regards the one-to-one correspondence between spatial-temporal activity patterns and the instances of mind postulated at the beginning of this chapter. Are the patterns *themselves* directly coding for the mind? In other words does the "same" pattern in two distinct brains code for identical mental states? This question is

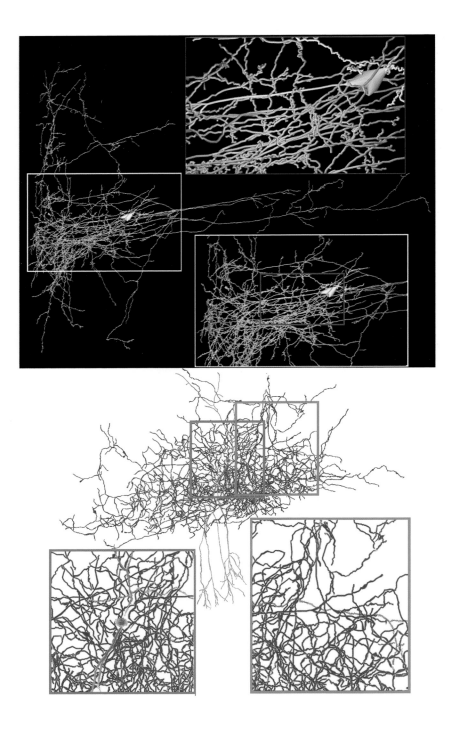

Figure 4.4
Neural forestry. (Top) A pyramidal cell from the mouse visual cortex.[18] The soma and selected dendrites are colored silver, the axon purple with yellow bifurcation points. The white and red insets display progressively more zoomed-in details of the arborization (rendering by Amina Zafar in the author's lab). (Bottom) A basket cell from the rat dentate gyrus.[19] Soma and dendrites are colored magenta, axon blue. Left and right insets zoom in near the soma and on distal arbors, respectively (rendering by Namra Ansari in the author's lab). Both reconstructions are freely available online at NeuroMorpho.org (branch thickness was increased to enhance contrast).

trickier than it seems because no two brains are identical. The activity patterns are identified not only by which neurons are firing and when but also by which neurons are *not* firing at a given time (because if they fired, the mental state would be different). Therefore, defining "sameness" for activity patterns (and thus mental states) in different brains is nontrivial to say the least.[17] For example, if a group of students learn a new concept, the newly learned activity patterns in all their brains would not necessarily be similar. Thus, different activity patterns in different brains could have similar meaning as representing the same concept for both individuals. This would imply that the mind is associated with the *information* represented in the activity patterns rather than with the specific patterns themselves.

More generally, we are almost completely ignorant about the essential signatures that link firing patterns to mental states thereby providing a reciprocal map between the two. Instead of attempting to solve this major challenge in the brain-mind problem, we turn our attention to other central issues of neuroscience, computation, and cognition, which are no less filled with mystery. Some of these fascinating matters, it turns out, can be couched in terms of trees and forests (figure 4.4).

5 Learning from Experience

5.1 The Second Principle of the Brain-Mind Relationship

One of the most striking aspects of our cognitive life is that it continuously changes, not just because it represents a world that continuously changes but also because of changes in the representation itself. In other words we continuously *learn* as a consequence of our experience. Consider the last time you studied for a test. Reading the book one day allowed you to retrieve the content during the exam the next day. Although you learned the content by reading the book, learning is not the same as the experience of reading the book; most content one reads is rapidly forgotten. Instead, in this context learning consists of the ability to retrieve that content at a later time. Again, it's not the experience of retrieving that knowledge that constitutes learning; if the test had been canceled, you would still have learned that material just the same.

More generally, we define *learning* as the acquisition of the ability to experience a previously inaccessible mental state. By definition, after we learn a new concept or fact, we know the content of what we learned, which we had not known before. Since knowing something is the ability to experience the corresponding mental state, learning is the acquisition of that ability. By the same token, forgetting something consists of losing said ability for that given something.

If knowledge is primarily encoded in the synaptic connectivity of the network (the *connectome*), the neural correlate of learning must be a change in network connectivity. The formation or elimination of new synapses would enable the activation of new activity patterns instantiating new mental states. This notion can be phrased into the second principle of the brain-mind relationship:

Learning, meant as the acquisition of the potential to instantiate a previously unknown mental state, corresponds to the change of connections in circuits of neurons.

Note that there is no one-to-one correspondence between creation of synapses and learning on the one hand and elimination of synapses and forgetting on the other. In appropriate circumstances learning might occur by disconnecting neurons. Likewise, forming new contacts among neurons might result in a loss of the ability to instantiate certain activity patterns and thus the related mental states. It is certainly plausible that learning would typically require concerted formation and elimination of synapses. These processes might occur together or even sequentially. Recent evidence, for instance, suggests that the number of synapses tends to increase during waking hours and to decrease during sleep.[1] One theory posits that synapses are continuously formed in random fashion between pairs of neurons, but only those that get activated become stable, and the others soon disconnect[2] (figure 5.1).

Creating or stabilizing synapses requires an experience (e.g., episodic event, mental association, being taught new facts). These are one-time, fast occurrences (seconds to minutes). How do synapses form (or fade) in such a short time span? Experimental evidence is now mounting to suggest that both the presynaptic (axonal) and postsynaptic (dendritic) specialized processes possess appropriate plastic properties to enable structural flexibility. In particular, dendritic spines can twitch in and out of the membrane shaft as well as wiggle about. In the meantime axonal varicosities can crawl up or down their branch in addition to occasionally getting absorbed by the axon, perhaps only to bubble up again at another time or location. Therefore, synapses might form or be eliminated by the movement of the spines, varicosities, or both.

Interestingly, the temporal scale of this bubbling and wiggling is exactly in the second-to-minute range, corresponding to the time necessary for learning something new. We expand on the temporal scale of neural plasticity and corresponding cognitive processes at the end of chapter 6, after depicting a more complete canvas of the relationship between the brain and the mind as well as of their changes with time.

For now, we ought to close our description of the second principle of the brain-mind relationship with one additional observation. The ability of day-to-day experience to form and eliminate synapses in the brain is not

Figure 5.1
Learning from experience I-II-III (Daniel Segrè, 2014, Halls Pond Sanctuary, Brookline, Massachusetts).

uniform throughout life in all of the nervous system. Some brain regions only maintain this capacity for a limited period during development, and their connectivity becomes largely fixed in the adult brain. Young children, in contrast, have far more malleable brains.

Consider, for example, the visual pathway. Neurons from the retina at the back of the eye[3] extend their axons all the way to specific "relay" neurons in an intermediate station in the middle of the brain (the thalamus). The axons of these relay neurons reach all the way to the visual cortex in the back of the head. In newborn babies the axonal branches of these retinal and thalamic neurons (and the corresponding postsynaptic dendrites) continuously fine-tune their positions until the appropriate neurons from every position of the retinas in each eye form indirect connections (via the thalamic relay) with neurons in the correct locations of the visual cortex, as described in section 4.3. Neurons from each tiny patch in the retina, say the bottom left corner, connect (through one intermediate synapse) with neurons in the corresponding "visual fields" of the neocortex; neurons from the right and left eyes find their partners in alternating ocular dominance stripes in the cortex; and color-sensitive retinal neurons match up with cortical blob neurons. The connectivity of this circuit remains plastic for only a short period after eye opening in infants, and after that the overall wiring of the visual network remains largely unchanged for the rest of one's life.[4]

Similar temporal windows of developmental plasticity (known as *critical periods*) exist for most sensory and motor modalities and the corresponding parts of the cortex. Moreover, these critical periods are not absolute in the sense that some residual plasticity persists later in life, but changes in neural structure require much more drastic interventions than during the critical periods. For instance, after a finger is amputated, nearby fingers eventually become innervated by the neurons originally responsible for touch sensitivity and muscle movement of the missing digit, but this adaptation may take years.

Even the regions responsible for understanding and producing language (named Wernicke's and Broca's areas, respectively, after the names of their discoverers) demonstrate these critical periods for learning. People who learn a foreign language during the age of three to eight years old usually grow as "native" (or bi-/multilingual) speakers as adults.[5] In these individuals, Wernicke's and Broca's areas are fully active when they speak or listen to that foreign language. Most people can learn new languages directly as adults, but they typically maintain an accent and have difficulty "thinking" (or dreaming) in the new language. Their Wernicke's and Broca's areas are much less active when their "owners" communicate in the adult-learned languages.

In other parts of the brain, especially the hippocampus and, to a certain extent, the prefrontal cortex, the critical period appears to be the entire life span. Interestingly, these brain regions are responsible for consolidation of memory and abstract reasoning, two cognitive skills that are involved in continuous daily learning of new facts, episodes, and concepts. This temporal and spatial selectivity of structural plasticity in the brain is perfectly tuned for an optimal balance between stability and flexibility of the corresponding functions. In particular, the life-long plasticity of hippocampus and prefrontal cortex allows adults to learn new abstract concepts and facts at an amazingly rapid speed even if they contrast with long-held beliefs.[6] In contrast, the limited critical period of the visual cortex is well suited to store and maintain forever the capabilities, learned in infancy, to see points, boundaries, faces, distances, contrast, and colors. We reconsider some of these important issues in chapter 7.

5.2 Probability versus Capability of Experiencing a Mental State

The notion that learning corresponds to synaptic formation or elimination is appealing, logical, and powerful (figure 5.2). Nonetheless, it is not (yet) commonly accepted in the research community.

Earlier we surmised that the first principle of the brain-mind relationship, linking mental states with neuronal firing patterns, meets the agreement (at least implicitly) of most active neuroscientists and maybe even an overwhelming majority if stated in its most general form, that electric activity, rather than spikes, encodes for the mind. In contrast, acceptance of the second principle, equating learning with a structural rearrangement of network connectivity, meets a certain level of resistance among brain scholars. One reason for this skepticism is that, although increasingly solid, the empirical proof of structural plasticity in synaptic connections is quite recent[8] and still known to only a minority of researchers in neurobiology. It is not yet widely described in textbooks, and it may take some time before it becomes mainstream.

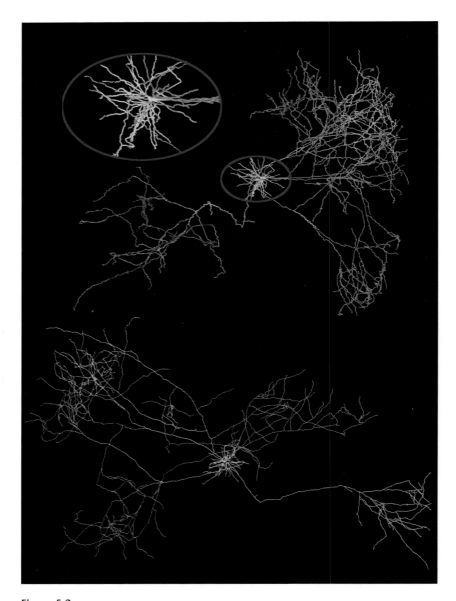

Figure 5.2
Networking to learn. (Top) A layer 5/6 pyramidal neuron from the cat visual cortex.[7] Axons are colored silver-blue, the apical dendrites purple, the soma and basal dendrites brown. The inset zooms in on the arbors near the soma. (Bottom) The same neuron rotated perpendicularly (view from the top of the apical tree) to visualize the span of the axonal subtrees (renderings by Namra Ansari in the author's lab). The reconstruction is freely available online at NeuroMorpho.org (branch thickness was increased to enhance contrast).

A perhaps more important reason that the second principle constitutes a minority view in neuroscience is the long-standing belief in another theory. According to a popular idea, already mentioned in section 3.3, learning corresponds to a change in the *strength* of synapses rather than their presence or absence. In other words the correspondence between learning and synaptic plasticity is commonly taken to occur at a "softer" level, in which it is not the *structure* of synapses that is plastic but rather their effectiveness in allowing communication between two neurons. In the jargon of artificial neural networks, what varies with learning is the synaptic *weight*, while the connectivity map among neurons remains constant. From the mathematical point of view, formation or elimination of synapses could be viewed as a special case of changing synaptic strength, as one could imagine assigning a weight of zero to nonexisting synapses.

Yet the functional and biological difference between the two theories is substantial. If a mental state corresponds to an activity pattern in the network, changing the connectivity of the network clearly can block or enable its instantiation. Solely altering the strength of the connections without adding or eliminating any of them, in contrast, could certainly facilitate or hamper the instantiation of an activity pattern and corresponding mental state but might not completely dial this possibility to zero. By analogy, adding a stop sign on a busy street would slow down traffic in a given direction, making it less likely for a driver to select that route. Nonetheless, only closing the road would ensure that the given traffic pattern is completely prevented.

We purport that synaptic weights constitute the neural correlate of the *probability* (rather than the *capability*) to experience a mental state. This distinction applies to our memories as well. We may be more likely to remember an autobiographic episode of the past if the synapses among the underlying neurons are strong. However, weaker synaptic connections will not wipe out the storage of a particular memory altogether.

At the same time, changes in synaptic strength and changes in synaptic presence are not completely independent processes. Consider for example a simple circuit of three nearby neurons in which the first is weakly connected to the second and the second to the third, but the first is (initially) not directly connected to the third, and there are no backward connections. Let's assume that the first neuron is spontaneously and randomly active, whereas the second and third neurons only respond to their inputs.

Because the synaptic connections are weak, it is possible but unlikely that a spike in the first neuron would trigger a spike in the second one and that the latter would make the third neuron fire. It is impossible for a spike in the first neuron to elicit one in the third without also causing the second neuron to fire as well. In other words this circuit does not "know" the "mental state" corresponding to the firing of the first and third neurons while the second is silent.

Now imagine that a casually rapid sequence of spikes in the first neuron systematically activates the second neuron so as to strengthen the synaptic connection from the former to the latter. Consequently, any spike in the first neuron would be likely to trigger a spike in the second. At the next random episode of bursting in the first neuron, the second neuron will also respond with many consecutive spikes, thereby making the third neuron fire and reinforcing the synapse between those two neurons. Now, with two strong connections, most spikes in the first neuron will result in a spike in the third neuron. This may cause the formation of a new synapse connecting the axon of the first neuron directly to the dendrite of the third neuron. At that point even if the second neuron dies, the circuit would have acquired the capability of implementing the activity pattern consisting of a spike in the first neuron immediately followed by a spike in the third one.

Correspondingly, variations in the probability of instantiating a mental state may affect the capability of experiencing another one. Suppose that you *know* the name of a famous singer, but you're not a particular fan of her music. Therefore, you *can* think of her name but seldom *do*. Now imagine that your best friend becomes obsessed with that pop star's latest hit, and you end up hearing the same songs over and over. You'd likely find yourself thinking of the singer's name much more often than before, and you'd soon develop a sense of familiarity (perhaps even boredom) that you didn't know before. The increased probability of experiencing a mental state (naming the singer in your mind) triggered the novel capability of instantiating a previously inaccessible feeling ("here she is again").

Strengthening and forming synaptic connections are therefore distinct but not completely independent phenomena, and just the same holds true for the probability and possibility of a conscious experience. Although the parallels between synaptic strength and experience probability, and between synaptic presence and experience possibility, are conceptually useful, the correspondence is not always so straightforward. Sometimes synaptic strength actually influences our *ability* to have an experience. Recall from section 3.2 that neurons continuously poll all synaptic signals throughout their dendritic arbors to determine whether or not to fire a spike down the axon at any one time. Now imagine two groups of neurons representing distinct concepts (for instance, "trees" and "brain") being interconnected by ten very weak synapses. Even if all the neurons of one group are active, their collective input might still be too feeble to make a difference as to whether the second group gets activated or not. However, strengthening these same ten synapses (perhaps by reading this book) would make it possible for one group to activate the other. In this scenario a change in the weight of existing connections effectively corresponds to learning the association between two concepts.

In the above example although the increased synaptic weight enabled the instantiation of the mental state "trees + brain," this experience remains infrequent because it requires all ten synapses to be active at the same time. Suppose you are reading the book again, resulting in the formation of twenty additional synapses between these two groups of neurons. You still need ten synapses to be active for one group to turn the other on, but these can now be any ten of the thirty available connections. You haven't really *learned* anything *new* because you already had the capability of retrieving the notion of brain trees, but this thought is now much more likely to occur. In this case synaptic formation altered the probability, rather than the possibility, of instantiating a particular neural pattern.

The above examples represent plausible but rather uncommon circumstances. The conceptual mappings of synaptic connectivity onto knowledge, and of synaptic strength onto the spontaneity of a thought, remain a convenient rule of thumb that we continue to follow in the rest of this book. Nevertheless, it is more reasonable to assume that specific combinations of both synaptic connectivity and strengths determine (with quantitatively different effects) both the capability and probability of experiencing a mental state.[9]

5.3 Panta Rhei

Long ago in ancient Greece, the pre-Socratic philosopher Heraclitus may have been onto something when he claimed that "everything is changing" (in Greek: *panta rhei*). Nobody can bathe twice in the same river, he argued, as the water flow makes the river different at every instant. Network connectivity in our brains may be similarly in flux, and correspondingly nothing remains still in the mind. We continuously experience mental states, and with each experience the network of our brain changes a bit. Even when we do not have a conscious experience, we might still adapt to "subthreshold" stimuli, as structural plasticity of dendritic spines can be triggered even by local electric activity that does not propagate throughout the whole neuron.

More specifically, billions of spikes crisscross the enormous axonal forest of our brain every instant. Each of these spikes evokes tens of thousands of tiny electric discharges on the other side of just as many synapses. As a result, we remember our first date, or we plan our next trip to the grocery store; we have a casual conversation with a neighbor, or we play the piano; we run as fast as we can to catch the bus, or we gently steer the driving wheel to keep the car in the right lane; we retract a burnt finger in pain or ecstatically savor a delicious dessert. But also thanks to those billions of spikes and trillions of synaptic currents our heart beats, our lungs breathe, our eyelids blink, our throat swallows, our stomach digests, and countless other muscles contract, relax, or twitch; our liver, kidneys, thyroid, and like glands release hundreds of chemicals into our bloodstream in finely balanced ratios; and of course the next ocean of spikes readies for discharge an instant later.

Yet the most consequential outcome of the brain's instantaneous spiking activity is not our present mental state, muscular activity, hormonal release, or electric discharge. The greatest effect is rather on the structure of the brain itself. The *plasticity* of neurons and synapses that we described in sections 3.3 and 5.1 is not just an optional, if intriguing, feature of nervous systems. It is, instead, an omnipresent staple that unavoidably accompanies all bits of activity. As trains of spikes travel down the axons and dendrites integrate synaptic information, billions of new synapses are formed, billions are wiped out, and the strength of billions other existing synapses adjusts up or down.

Therefore, the relationship between structure and activity in the brain is fundamentally one of *reciprocal* cause and effect. The connectivity of the network along with all synaptic weights determines what patterns of activity can be instantiated and which among those are selected at any time.

Conversely, the continuous flow of activity steadily sculpts and resculpts the connectome and its synaptic weights from before birth to death. Imagine a gigantic system of highways in which the passage of every car alters one way or the other the number and width of the lanes, the speed limit, and even which exits are open or closed, and where they are. That's each cubic inch of your cerebral cortex, day and night. This means that both the *probability* of instantiating a given mental state as well as the *capability* to do so (i.e., knowledge) can change continuously as an individual goes about her or his experience (figure 5.3).

Then again, if neurons are so plastic, how can they store memories for as long as one hundred years? If all neurons continuously alter their synaptic connectivity, why don't we risk losing really important memories in the course of everyday experience? Experiments tracking dendritic spines in vivo during normal animal behavior show that spine twitching may be a fairly rare event. In particular, in the mouse cortex any one synapse appears to remain stable for about a month.[10] But how often would we *expect* synapses to form or prune? It turns out that even the very low observed rate of structural plasticity is amply sufficient to underlie learning. Let's suppose just for the sake of argument that at every moment of "mental time" (50 milliseconds), exactly one synapse turns over somewhere in the nervous system. Of course learning requires more than just one synapse to change, but learning something new also typically takes more than 50 milliseconds. Then in a month only approximately 50 million synapses (30 days times 24 hours per day, times 3600 seconds per hour, times 20 mental instants per second) would twitch. Of 10^{12} connections in a mouse brain, this would give us a chance of just 0.005% to catch an individual synapse in the act! This simple calculation illustrates how challenging the experimental lab work can be in this field.

The above reasoning demonstrates the compatibility between the second principle of the brain-mind relationship and the relative structural stability observed experimentally. Moreover, the incredible double-dipping achievement of life-lasting memories and hourly learning is essentially enabled by the huge numbers of neurons and synapses. A scaled-down network would need to choose between slow memory decay and fast learning, but a brain with trillions of synapses can afford both.

At the same time, there appears to be a balance between the rate of learning and the rate of forgetting because both learning and forgetting are

Figure 5.3
Mind the trees. (Top) Walking into a new mental state (Daniel Segrè, 2014, Fresh Pond Park, Cambridge, Massachusetts). (Bottom) Trees in a modified mental state (Daniel Segrè, 2014, Fresh Pond Park, Cambridge, Massachusetts).

mediated by changes in synaptic circuitry. The quicker synaptic turnover is, the faster the circuit can store novel information, but also the faster it is bound to forget previous data. The brain, however, evolved a clever way out of this conundrum by setting different rates of synapse formation and pruning in separate areas of the nervous system. One specialized cortical region, called the hippocampus, has faster synaptic turnover (but less storage capacity) than the rest of the cerebral cortex. The hippocampus serves the role of "fast learner," while the rest of the neocortex takes care of steady storage.

How can this division of labor between specialized brain structures result in long-lasting memories in the face of fast learning? We can understand the trick using a simplified cartoon. Imagine that every neocortical neuron is reciprocally connected with a corresponding neuron in the hippocampus. Examples of hippocampal neurons that are known to make synapses with neocortical neurons are shown in figure 5.4. We describe this network in section 8.3, but the circuitry details are not important here. Consider two stimuli, A and B, represented respectively by neurons a and b in the hippocampus and by corresponding neurons α (alpha) and β (beta) in the neocortex. After only a few A-B co-occurrences, a and b will form a synapse (because they have fast structural plasticity), but α and β will not (due to their slower structural plasticity). Now every time A is observed even in the absence of B (or vice versa), both a and b will fire, causing α and β to fire together as well. Eventually, α and β will also form a synapse. Soon enough, if A continues to be observed without B (or vice versa), a and b will get disconnected (because of their faster structural plasticity), but the connection between α and β will remain long lasting.

Now pretend that A is your first "significant other," and B is the first kiss. That first kiss was rapidly and temporarily stored by a synapse between neurons a and b in the hippocampus. In the course of a month of dating, you saw A many times, and the memory of that first kiss was eventually "consolidated" into a synapse between neurons α and β in the neocortex. However, after breaking up, you still saw A many times (unfortunately with no more kissing), and hippocampal neurons a and b got disconnected. Nevertheless, thinking of A many years later, you can still remember the first kiss (through activation of neuron β). Clearly, this example should not be taken literally, not the least because stimuli, significant others, and kisses

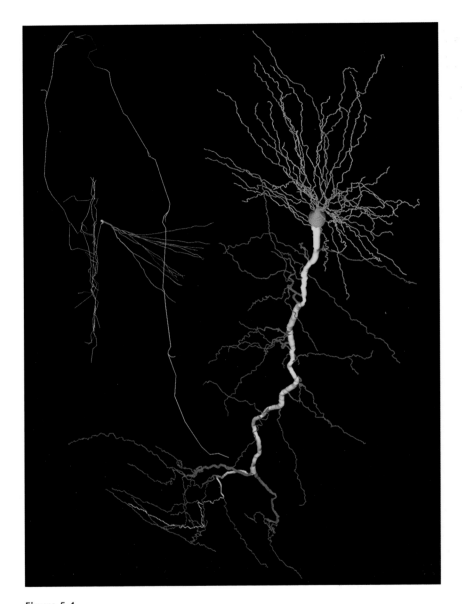

Figure 5.4

In and out of the hippocampus. (Left) A granule cell from the dentate gyrus of the rat hippocampus (the "entry" neuron of this circuit). The dendrites[11] and axons[12] are colored deep and light blue, respectively. (Right) A pyramidal cell from region CA1 of the rat hippocampus[13] (the "exit" neuron of this circuit). The basal dendrites are colored light blue and green, the apical trunk is yellow with oblique branches red (renderings by Todd Gillette in the author's lab). Both reconstructions are freely available online at NeuroMorpho.org (branch thickness was increased to enhance contrast).

are not represented by individual spikes in single neurons but (in the neocortex and hippocampus alike) by neural patterns in cell assemblies.

Even with this kind of differential plasticity in separate brain areas, it is unlikely that we literally learn something in every moment. Only some experiences trigger learning. It is feasible, however, that every mental state does in fact change the subsequent probability of experiencing other mental states. In other words, although synapse formation and elimination might happen in only some circumstances, our brains continuously adapt at least in terms of synaptic strength.

Yet simply stating that neurons form new synapses "sometimes" is not quite satisfying. When and why does or doesn't synaptic circuitry change? Why do we blank when facing the spouse of our spouse's best friend, as we once again forget his or her name just a year after the last holiday party together? Why don't we ace each and every school exam after a night spent cramming? Why do we only have a foggy recollection of playing with grandma when we were toddlers?

5.4 To Learn or Not to Learn

We all forget much of what we learn. For example, we can only retrieve a very small portion of all school material we absorbed at the time, and that goes even for straight-A students. With a bit of time and concentration, we can think of a few birthday gifts we received through our childhood, when we surely could have effortlessly listed each and every one a week after each birthday. This is all knowledge that we did in fact learn but just as easily forgot with the passing of time. What color socks was I wearing two weeks ago on Wednesday? Was there snow on the ground last New Year's eve? What is your niece's favorite ice cream flavor? Suitable synapses were in place in the brain to store this knowledge at some earlier time, but they since succumbed to rewiring.

Besides the inevitable forgetting of knowledge we did once have, however, there is also a huge amount of potential knowledge that we are constantly exposed to, yet we just never learn. Indeed, it seems that we only learn (i.e., acquire the ability to later remember) a minute fraction of what *could* be learned.

There are many events that happen around us and that we fail to even notice: a bumper scratch on the car in front of us at a stop light, a distant gray cloud in the blue sky, and the soft background music in the waiting room of a doctor's office. We wouldn't expect to learn details that we don't even pay attention to in the first place. But how about present information that we are clearly aware of, even responsive to[14]? How about conscious actions and perceptions that we experience not just once but repeatedly? Even among these facts and events, we only learn the tiniest of proportions!

I consider this partial capability of absorbing information a fortunate cognitive feature. Imagine if you could recall every state of mind you have ever experienced.[15] Each sight, sound, smell, and taste, every intention, feeling, thought, all of your body's movement you were ever aware of. You'd run the serious risk of being flooded with useless information, and it might become challenging to find any meaning in such an ocean of knowledge.

Although certain characteristics of experience (e.g., strong emotions or consistent repetition) can make specific mental states particularly memorable, in most cases we do not have a deep or comprehensive understanding of the cognitive "signatures" that discriminate what we learn from what we do not. We can nevertheless safely accept that we are bound to learn only a small proportion of the potential knowledge we are continuously exposed to. Of course, learning something doesn't ensure that we will have an opportunity to retrieve that knowledge in the course of our life. Much of what we learn and know is available to our mind, but we never get to "use" it.

In section 4.2 we offered a (philosophically) "idealist" view of reality based on the notion that we experience only a small fraction of what we *could* experience. In the same vein if we start from the standard (*materialist*) assumption of a "world out there" in which events happen, we can develop a (more scientifically traditional) materialistic view of reality based on our limited or selected learning ability:

Many episodes could happen

A few of those do happen

Each person observes a subset of the episodes that happen

A few of the observed episodes are learned

A fraction of the learned episodes are later remembered

Figure 5.5
Branching arbors, memory's crucial ingredient (photograph by Daniel Segrè).

This materialist view of reality, in which mental states *result from* occurrences in the world, and the idealistic viewpoint presented in the last chapter, in which mental states *result in* occurrences in the world, are not incompatible with each other. Instead, they represent complementary perspectives on the relationship between mind and matter.

Where does this leave us in terms of physical underpinning of learning and memory? How does the fact that we are capable of learning a few of the events we observe but not the majority of our experiences reflect the underlying brain mechanisms of our mental existence? In other words, can we identify appropriate neural correlates of this mysterious facet of our cognition?

The last chapter presented the idea that neural activity (in the form of spiking patterns) corresponds to mental states. In this chapter we argued that learning corresponds to formation and elimination of synaptic connections, and thus the entire connectome of an individual brain constitutes that person's "knowledge." We mentioned mounting scientific evidence that supports such activity-dependent rewiring.

If brains are reconfigured through experience, when or why do we learn something, and when or why do we not? Somehow synapse formation must require more than just activity (figure 5.5). Experiencing a mental state may well be necessary for creating the ability to retrieve it later, but it is by no means sufficient. Might the tree-like branching patterns of axons and dendrites provide the missing piece of this puzzle?

6 The Capability of Acquiring New Knowledge

6.1 The Third Principle of the Brain-Mind Relationship

We are at a crucial passage of our exposition, and it is worth briefly recapping the main elements of the conceptual architecture erected so far. The human brain is a gargantuan network of one hundred billion neurons. Each of these neurons is endowed with two complementary kinds of tree-shaped extensions: dendrites and axons. Dendrites are computationally sophisticated devices that invade a volume thousands of times the space occupied by the neuron's soma (cell body) to integrate input from tens of thousands of other neurons in the form of graded electric signals. Dendrites elaborate this information through complex nonlinear biophysical dynamics and convey the results to the soma. Axons are hyperextensive branching cables that faithfully and rapidly deliver the neuron's output to tens of thousands of other neurons throughout the entire brain. Axons encode their messages in series of binary (all-or-none) spikes, each lasting as little as a millisecond. Both axons and dendrites operate in massively parallel fashion, with hundreds-to-thousands of branches in every neuron independently carrying out their duties at once. Mental states correspond to patterns of activity distributed in space and time over a large number of neurons (the first principle of the brain-mind relationship).

The elementary communication in the nervous system occurs at each of the thousands of trillions of brain connections between the axon of a neuron and the dendrite of another. These synapses are highly plastic, not only because their strength changes with every impulse they transmit but also because they can appear and disappear. Although a very small fraction of synapses turns over at any one time, the total number of synapses in the brain is so huge that the count of connections that change with each experience is still staggering. Knowledge is the ability to achieve a certain mental state; the capacity of the brain to instantiate the corresponding activity pattern resides in its synaptic connectivity. Thus, learning (and forgetting)—the acquisition (and loss) of knowledge—correspond to changes in the synaptic circuit of the brain or connectome (the second principle of the brain-mind relationship).

Can these foundations reveal the neural underpinnings of why human beings effortlessly learn essential new facts from certain experiences and not from others? If forming new synapses is a critical correlate of learning, we should focus our attention onto the necessary conditions for synaptic generation. In chapter 3 we explained that if two (connected) neurons are activated at the same time, their synapse becomes stronger; this is what some neuroscientists call "Hebbian" plasticity. The last chapter introduced a form of experience-dependent structural plasticity in which a synapse can be created between two neurons that were not previously connected. By putting together these two types of plasticity we can understand the meaning of a popular adage among neuroscience researchers: neurons that "fire together, wire together"; *this* is what *other* neuroscientists call "Hebbian" plasticity. In other words if two neurons get consistently coactivated, they are likely to end up synaptically connected.

The "fire together, wire together" notion has an appealing cognitive correlate: associative learning. We tend to recall together thoughts or events that occurred together. For example, after seeing the face of our office mate every day for years, we recognize her even when the view is partially covered by a computer monitor, obstructing the mouth, nose, and the bottom of the eyes, and only revealing the eyebrows, forehead, and hair. In our mind, all of her face parts are tightly associated together, so that the "missing parts" get seamlessly filled in with the available information.

This cognitive skill, known as *pattern completion*, can be explained computationally by the "fire together, wire together" mechanism: by consistently firing together every time we see our office mate in full view, the neurons representing all parts of her face wire up together. When we see the face behind the computer monitor, the visual signal from the retinas at the back of our eyes can only activate some of these neurons (say, those coding for the hair and eyebrows). However, the synaptic connections of these active neurons with the neurons representing the rest of the face are sufficient to trigger the activity that was not directly caused by the partly occluded sight. As a result, even though our *retinas* only "see" hair and eyebrows, our *mind* "sees" the entire face.

Yet when the brain is envisioned as a gigantic forest of axonal and dendritic arbors, it becomes clear that firing together cannot possibly be sufficient for wiring together. A basic additional requirement for an axon to make contact onto a dendrite is for some of their branches to share the same space! In other words a new synapse can only form between two neurons if the axon of the (would-be) presynaptic neuron passes in close proximity to a dendrite of the intended postsynaptic neuron. If the axon of a neuron is nowhere near the dendrite of another neuron, you can make those two neurons fire in sync all you want, but they will remain physically unable to form a synapse. Although such a constraint is so reasonable as to appear obvious,[1] it allows us to draw a very nontrivial conclusion, which may well represent the central tenet of this book. This is the third principle of the brain-mind relationship:

Axonal-dendritic overlaps are the neural correlate of the capability of learning.

Those tree-shaped structures that receive, process, and transmit signals are not just the means nature found to create the most complex networks we know. They don't simply enable the flow of information and computation in the brain. The very branching structure of those axonal and dendritic arbors dictates not only which pairs of neurons are connected by synapses, but even more consequentially it dictates which pairs of neurons <u>can possibly</u> wire together (that is, if appropriate experience makes them fire together).

For this reason, axonal-dendritic overlaps are sometime referred to as *potential* synapses.[2] Stated differently, a spatial proximity between the axonal branch of one neuron and the dendritic branch of another neuron constitutes an opportunity for that pair of neurons to form a new synaptic connection. Because changing the connectivity of the circuit corresponds to alteration of knowledge (e.g., memory formation), the relative positions of axons and dendrites can determine how experience may or may not shape our memory. After all, these axons and dendrites provide a possible explanation for an intriguing yet familiar aspect of our mind: why we learn and remember certain things, while others we do not. Axons and dendrites represent, I like to think, the actual *tree of knowledge* of biblical reference, as quoted at the beginning of chapter 1.

The visual metaphor of axons and dendrites as trees is extremely powerful, so much so that this entire book wrote itself around such a floral view. The attentive reader, however, may have already noticed a hard limit to the parallel between neuronal trees and actual vegetation. Whereas the job of axonal and dendritic branches is to connect to each other, every plant in a garden is by and large physically isolated. To put it bluntly, forests are not (neural) networks.[3] Nevertheless, botanical trees are again useful to understand axonal-dendritic overlaps in concrete terms. Imagine a lush wood dense with trees whose trunks sit only feet apart. As you daydream your way into the woodland, you notice that the ground is completely shaded. When looking up, you see a thick tapestry of intersecting branches, so complex and impenetrable that it is hard to say which tree any one branch belongs to. Although those arbors do not make synapses with one another, their leaves might nearly touch each other (figure 6.1). Every "close encounter" in which the branch of one tree passes within a leaf length of the branch of another tree may serve as a visual aid to imagine an axonal-dendritic overlap. The leaves can be considered as stand-ins for axonal varicosities or dendritic spines.

Now remember from section 2.2 that dendrites and (especially) axons have a much more slender shape than "real" arbors: at the scale of a typical oak tree, the trunk of an axon would be as thin as your pinky finger. Conversely, if we were to expand an axonal trunk to the size of that of an oak tree, its limbs would reach many thousands of miles away! Furthermore, no matter how dense you imagined your forest, the brain is denser. There is no space, no "air," between trunks, branches, and leaves in the nervous system, no narrow path for a shrunk reader to walk through: the entire volume is filled. These differences between neuronal and botanical trees (the former being much thinner, longer, and more densely packed) make apparent that axonal-dendritic overlaps are very, very common. While looking up in your imaginary botanical forest you could see just a few hundred juxtaposed branches; from the same position in the brain you'd see tens of thousands. Indeed, you may recall the overwhelming number of synapses in the human brain. There are even more axonal-dendritic overlaps than synapses, since you cannot have the latter without the former. Moreover, axons can extend so far as to traverse the entire brain from side to side. In your imaginary woodland, two trees from different continents or from opposite Earth hemispheres would need to stretch over the span of

The Capability of Acquiring New Knowledge

Figure 6.1
Their leaves might nearly touch each other (Daniel Segrè, 2014, Fresh Pond Park, Cambridge, Massachusetts).

billions of other trees in order to overlap their respective branches. Neurons in the brain do precisely that.

Yet, just as every neuron makes synapses with only a minuscule fraction of all other neurons, the same occurs for axonal-dendritic overlaps. As you may recall from chapter 2, a human brain is estimated to have a little less than 100 billion (10^{11}) neurons and a quadrillion (10^{15}) synapses. Based on available experimental evidence, we can reasonably estimate that there are 10 quadrillion (10^{16}) axonal-dendritic overlaps.[4] These numbers are indeed very large but "only" amount to an average of 100,000 overlaps per neuron. This means that the axon of a typical neuron comes in close proximity of the dendrites of just one of every million other neurons[5]—that is 1% of 1% of 1%. Nevertheless, two neurons, A and B, without an axonal-dendritic overlap have a much better chance of being "indirectly" overlapping with a third neuron C, such that the axon of the first neuron (A) overlaps with the dendrite of the "intermediary" neuron C, and the axon of this neuron C overlaps with the dendrite of B.[6] Any two neurons are virtually guaranteed to have indirect axonal-dendritic overlaps when two or three intermediate passages are allowed. This concept might be best understood in terms of the "degrees of separation" parlor game. Any two Hollywood actors or actresses may not have (yet) worked together on any movie, but in many cases a third famous actor (such as Kevin Bacon) can be found who has acted with them both.

To summarize, the capability of making new synapses can be understood in terms of physical proximity between the axons and dendrites of the respective presynaptic and postsynaptic neurons. As so often happens in biology, however, the picture is a bit more intricate than intuition would suggest. There isn't a one-to-one correspondence between thoughts, concepts, or feelings and individual neurons (let alone axonal and dendritic branches). Instead, even what we might consider unitary elements of a mental state—a color, an urge, or a fleeting impression—are each represented by "cell assemblies" consisting of complex activation patterns of a substantial number of neurons. Thus, just a simple association of a sound to a sight (for instance, that of a buzzing beetle) likely requires the concerted formation of multiple synapses and therefore the spatial overlap of all the corresponding neuronal cables.

These points notwithstanding, the story that emerges from the third principle of the brain-mind relationship becomes more interesting as we

The Capability of Acquiring New Knowledge

dig deeper. We are about to find out that this same organization explains other apparent mysteries of the human mind. Most intriguingly, this unique neural architecture also endows our cognitive machinery with superior computational powers.

6.2 Learning Is Gated by Background Information

After briefly glancing at a new music score, a proficient pianist quickly learns to play the piece and can hum it at once in her mind. Her beginner piano students, examining the exact same score, neither learn to play it nor are able to imagine the tune. Similarly, on reading for the first time the proof that the square root of two cannot be expressed as a ratio of integer numbers, a math-savvy college graduate remembers the theorem demonstration for years. Without basic math, his twin brother reading the same material forgets it in a few hours. In a third scenario, noting a new stop sign in your neighborhood once on the way back home one day, you would certainly recall it for several days. Seeing a new stop sign in a different intersection you only seldom drive through may not be sufficient to cause you to remember it a few days later. Many other examples like these can be conjured: it's often easier to learn about matters that we already know a fair amount of than to acquire the same knowledge starting from scratch. This familiar sensation can be expressed as the notion that *learning is gated by background information*. In other words one has to know enough basics about a subject in order to quickly learn more about it.

Although this observation is corroborated by everyday experience,[7] it also appears to present a logical paradox. Why is one-trial learning often easier for children, even though adults boast more extensive background knowledge? Prior expertise may be a less stringent requirement early in life than it is for grown-ups. Kids appear to be capable of absorbing any new information even when they start almost entirely clueless about the topic at hand. One possible mechanism for this greater learning capacity of young people could be their ability to master the essential foundations of new domains of knowledge very quickly. Being an expert in a given field might only matter for learning in a relative sense, that is, compared to how knowledgeable one is in other fields. Newborn babies know next to nothing; therefore, on a relative scale they are also potential experts of everything.

For a related apparent conundrum, how can adults learn pretty much any truly new knowledge or skill "from scratch" when starting from limited if any experience in that domain (figure 6.2)? How can we acquire the relevant background knowledge when we are missing the necessary basics? A possible way out of this chicken-and-egg scenario between preexisting expertise and learning anew is to leverage the power of repetition. Background information is indeed necessary for one-trial learning, but even a novice can acquire new skills and knowledge with enough persistence in training.[8] Another way of stating the same idea is that contextual knowledge is not strictly required for learning, but it greatly accelerates the storage of new information.

These constraints can disappear in highly emotional or surprising scenarios. A single shocking event can be remembered for one's entire life without need for rehearsal and in the absence of any relevant prior experience. For example, if you have a car accident at the new stop sign in the unfamiliar neighborhood, you'd likely remember that intersection for a while. But here, let's set aside repetition and emotional valence to focus on the everyday experience of one-trial learning of emotionally neutral new facts.

Why should this most common form of learning depend on background information, and what does this have to do with the third principle of the brain-mind relationship? We know from the *second* principle that learning corresponds to the formation and elimination of synapses. If making a new synapse requires an axonal-dendritic overlap, and learning requires background information, could it just be that axonal-dendritic overlaps are the neural correlates of background information? This may sound like an outlandish proposition at first. However, when accounting for the tree-like structure of axons and dendrites, this hypothesis turns out to be far from unreasonable. In order to understand why this is the case, let's start from an oversimplified cartoon in which each concept or mental state is encoded by a single neuron (the so-called *grandmother cell* hypothesis, briefly described in section 4.1).

Specifically, let's presume that there is one neuron that codes for buzzing and another that codes for beetles. Suppose that on noticing a buzzing beetle for the first time, you learn that beetles can buzz. Under the grandmother cell hypothesis, this means that the axon of the neuron encoding for the buzzing sound formed a new synapse with the dendrite of the

Figure 6.2
Wonder branching. (Top) How can we learn anything? (Daniel Segrè, 2014, Fresh Pond Park, Cambridge, Massachusetts). (Bottom) Three trees (Daniel Segrè, 2014, Halls Pond Sanctuary, Brookline, Massachusetts).

neuron encoding for the sight of a beetle. Thus, the next time you hear buzzing you might think of a beetle because the action potential in the "buzz" neuron now triggers a postsynaptic depolarization in the "beetle" neuron that could make it spike. Before your lone experience of the buzzing beetle, the two neurons were not connected, and hearing a buzz would not have elicited the thought of a beetle (that is to say, you did not yet know that beetles could buzz). Without a synapse in place, an action potential in the "buzz" neurons would not affect the activity of the "beetle" neuron: the two neurons would behave independently.

The third principle of the brain-mind relationship tells us that, if a synapse could form between the "buzz" and the "beetle" neurons (making you capable of associating these two concepts), the axon of the former neuron must come very near the dendrite of the latter. Why would that be so? This is not likely to happen by chance: as we have just established in section 6.1, the axons of any given neuron can only be in close proximity with a very small fraction of dendritic trees from all the other neurons in the brain! The more general question is: why is an axon placed where it is, as opposed to any other location in the nervous system? To find the answer, recall that the brain is a giant network, and the principal function of axons is to make synapses with dendrites. Thus, if you find an axon somewhere, it must be because it is making a synaptic contact nearby. Now back to our original quest: why should the axon of the "buzz" neuron be adjacent to the dendrite of the "beetle" neuron? Following the above line of thought, the reason is that the "buzz" axon is connected to other neurons that are located just next to the beetle neuron.

The key questions then become, which neurons is "buzz" contacting, and why are they sitting beside "beetle"? The first question is easy: what animals did you already know that buzz? Likely these include wasps, bees, and flies. The next issue is less trivial: why are the neurons encoding for all of these creatures grouped together in the same brain region? The reason is the same: circuitry. These neurons are near each other because they all receive synapses from the same sets of (other) axons! For instance, the above-mentioned insects are all connected with the "erratic flight" neuron and with the "crawling" neuron, just to mention a couple. If all the neurons representing insects were scattered randomly across the brain, the axons representing their shared features would need to travel even farther distances to connect them all. Instead, geographical proximity of related

The Capability of Acquiring New Knowledge

neural representations is an optimal design to minimize axonal wiring. Ultimately, this organization frees up some space to pack more neurons in the same volume, thus enabling greater information storage. So after all, axonal-dendritic overlaps actually reflect connections among neurons encoding related concepts, that is, precisely background information! Thus, the architectural principle linking axonal-dendritic overlaps with the existing circuit connectivity[9] constitutes the neural correlate of background information–gated learning.

To summarize, the existence of synaptic connections among neurons corresponds to knowledge, that is, the ability to have a particular mental state. The strength of these synapses reflects the probability of activating the thought corresponding to the underlying neural pathway. The spatial overlap between specific pairs of axons and dendrites provides the opportunity to form new synapses, which is to say, the ability to learn (figure 6.3).

Let's try to reduce these somewhat theoretical ideas to practical everyday life scenarios in the domain of spatial navigation. The task at hand is to find a reasonably short path between two locations. In these examples we describe the putative neural correlates in distinct fonts.

Having lived several years at your current address, you know fairly well the directions from the nearby parkway entrance to the large shopping mall 10 miles away, a route you typically drive every few months. Sometimes spontaneously (or whenever asked) you think of the directions, perhaps while planning to buy a gift for your best friend's birthday at the mall. In your brain synapses are present among the appropriate neurons corresponding to the knowledge of these directions. Their activation (that is, the discharge of the appropriate activity pattern in their neuronal assemblies) constitutes the instantiation of the thought of those directions. Your cousin visiting from out of town has never driven on his own from the parkway to the mall and has no clue how to arrive there without a GPS. His brain has never formed the necessary synapses to instantiate the activity patterns encoding for the thought of those proper directions.

After taking a new job in an office in that mall, you start driving that route almost every day and soon find yourself spontaneously thinking of those driving directions much more often. In your brain the frequent activation of the corresponding cell assemblies has

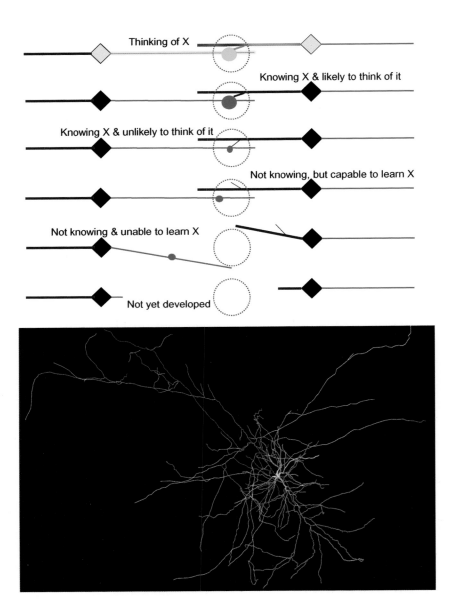

The Capability of Acquiring New Knowledge

Figure 6.3

The three principles of the brain-mind relationship. (Upper panel) Schematic summary of the neural correlates of mental states. Top (Thinking of X): A mental state, such as thinking of the letter X, corresponds to a pattern of activity represented here as a single spike flowing from the axon of a presynaptic neuron (orange) to the dendrite of a postsynaptic neuron (light blue) through the synapse formed by their varicosity (orange swelling) and spine (blue twig), respectively. In reality an activity pattern of even the simplest mental state involves numerous spikes flowing through a cell assembly of many neurons. Second from top (Knowing X & likely to think of it): Even when silent (red axon and blue dendrite), a circuit with strong synapses that could underlie the activity pattern corresponding to the thought of X (large axonal varicosity and thick dendritic spine) has a high probability of instantiating that mental state. Third from top (Knowing X & unlikely to think of it): An inactive circuit with weak synapses underlying the thought of X (small varicosity and thin spine) still possesses the ability, but only a low probability, to instantiate that mental state. Third from bottom (Not knowing, but capable of learning X): A circuit without the necessary synapses to spontaneously instantiate the thought of X (lack of contact between varicosity and spine) can acquire this capability by one-time experience of the corresponding activity pattern (e.g., firing of the left neuron followed by firing of the right neuron) if the appropriate axons and dendrites are in close proximity. Second from bottom (Not knowing & unable to learn X): A circuit without adequate axonal-dendritic overlaps to enable formation of the synapses necessary to mediate the thought of X by just minor flinching of varicosities and spines is unsuitable for one-trial learning of the corresponding mental state. Bottom (Not yet developed): Even the slower learning through consistent repetition, implying extensive neuronal rewiring, requires developed axonal and dendritic substrates. If these are not yet mature, no learning can occur. (Lower panel) A pyramidal neuron from layer 2/3 of the rat somatosensory cortex.[10] The dendrites are colored to emphasize branching complexity for contrasting the oversimplified cartoon of the upper panel (rendering by Todd Gillette in the author's lab). The reconstruction is freely available online at NeuroMorpho.org (branch thickness was increased to enhance contrast).

reinforced their synaptic connections, increasing the probability of activating that mental state.

One day a colleague of yours, who lives along the way, asks for a ride. Although you have never been to her house before, you easily learn the detour after being told only once how to get there. In your brain the cell assemblies corresponding to the directions to your colleague's house were not connected before she explained the route. However, the set of axons and dendrites of the neurons in these assemblies were already overlapping because you already knew many other details of the local geography: the axons and dendrites of the neurons coding for the route to your colleague's house were connected to other nearby neurons representing relevant background information. Their overlap enables the formation of new synapses when the corresponding neurons are coactivated by listening to the driving directions only once.

Your cousin visiting from out of town is present at the same conversation in which your colleague explains how to get to her house. However, without knowing much about the local road layout in the area, your cousin will not remember those new directions later on. In his brain the neurons coding for some aspects of those directions do get coactivated while listening to the explanation, but they cannot form synaptic connections because their axons and dendrites do not overlap, corresponding to the lack of relevant circuit connectivity coding for background information.

Although the above scenarios only constitute illustrative examples, the occurrence of similar situations (or a graded continuum of intermediate cases) can be readily recognized in almost every moment of our experience.

6.3   Learning the Truth

We have so far described how the third principle of the brain-mind relationship explains why appropriate knowledge of relevant background information gates the acquisition of new information by one-trial learning. The relationship between axonal-dendritic overlaps (corresponding to the potential to learn) and existing connectivity (representing knowledge) provides a direct neural mechanism for the familiar observation that experts can grasp new concepts in their discipline much faster than novices. Everyone finds it easier to learn new facts in their domain of expertise than in a completely novel field. This formulation, however, could be argued to be excessively optimistic. Those who perceive glasses half-empty rather than half-full might rather claim that background information gating works to hinder, not to facilitate, learning. In other words it isn't that experts learn better than novices. It is that novices learn worse than experts.

This may not necessarily be simply a matter of futile polemics over word choice. When we consider the neural correlate of this phenomenon, it is

The Capability of Acquiring New Knowledge

clear that requiring axonal-dendritic overlaps for synapse formation adds a constraint relative to a pure "fire together, wire together" plasticity rule. In a putative world unconstrained by physical requirements, it would be possible to postulate that if two (as-yet disconnected) neurons spike in sync, they could form a new synapse, independent of their preexisting connectivity. The third principle additionally demands that these two neurons already have an axonal-dendritic overlap in place for this to happen. Relative to the (simpler, but more naive) situation described before the beginning of this chapter, the added burden introduced by our "anatomical" considerations hampers learning. A network without the third principle would learn more, not less, than real brains.

Wouldn't it be wonderful if we could learn anything we encounter, without regard to what we know already? Before getting carried away with this delusion, let's remember the abnormal condition of hyperthymestic patients (mentioned at the end of the last chapter), who cannot help but remember nearly each and every episode of their existence. It could become harder, not easier, to find meaning in such an overcrowded mental repertoire. After all, the amount of time available to reminisce one's past is the same—independent of how many different memories are to be chosen from for retrieval. Even with a more limited set of learned associations (as constrained by relevant background information), the number of available memories far outstrips that of effectively recalled memories in the course of a lifetime. Learning too much information would make it more difficult to select interesting or relevant memories for later retrieval. Our thoughts would be constantly cluttered by irrelevant details of past episodes.

Even then, would our cognitive experience be more interesting if we could memorize the same small fraction (say 1%) of all our experiences independent of what we know already? What if a professional piano player and a math professor were on equal footing learning a new sonata or the latest theorem? Individually, such a hypothetical skill set would result in expanded cognitive breadth at the cost of depth. Everyone would become jacks of all trades and masters of none. Although the more superficially curious among us might like this scenario, the lack of learning bias introduced by axonal-dendritic overlap would likely reduce the emergence of special geniuses such as Mozart and Einstein. Imagine the loss for human society if young Wolfgang Amadeus had switched to studying physics after composing the first symphony! At the population level, cognitive differentiation is certainly an adaptive trait.

A bit of knowledge specialization may also provide an evolutionary advantage not just for societies at large but for the "average" individual as well. Imagine two human beings, one growing up in a tropical forest, the other in the arctic tundra. The former becomes an expert in recognizing poisonous insects, whereas the latter develops a particular acuity to spot caves at a distance for warmer sleep. It makes sense that it is advantageous for the person living in the tropics to learn ever more kinds of poisonous insects, forgoing the ability to recognize caves. For the folk stuck in the tundra, it is certainly more convenient to remember each and every cave seen in the recent past, even at the cost of remaining ignorant about insects. Thus, background information is typical of one's native or most habitual environment, and that is the most useful domain of knowledge for expertise refinement. This same line of reasoning also applies to other animal species and their habitats, not only humans.

In addition to all of the above arguments, however, there is a deeper and most compelling reason that makes background information–gated learning a very powerful computational mechanism. Specifically, *constraining synapse formation by axonal-dendritic overlaps makes it relatively easier to learn what is true than to learn what is false.*

Consider, for example, the scenario described a few pages ago, in which someone learns that beetles can buzz from a single observation of a buzzing beetle. When seeing the beetle while hearing the buzzing sound, this person probably experienced other sensations, for example, the savoring of a scoop of ice cream. The neurons (or rather the cell assemblies) representing the sight of the beetle, the buzzing sound, the taste of ice cream, the cold feeling on the tongue, and the texture of the spoon were all active together. Why should the brain learn that beetles can buzz and not, say, that a spoon can buzz or that seeing beetles should be accompanied by a sweet, cool feeling in the mouth? These seemingly absurd propositions are all consistent with a simple "fire together, wire together" rule of structural plasticity.

Classic learning theories that do not take neural branching overlaps into consideration explain that the "spoon" and "buzz" neurons, or related cell assemblies, do indeed strengthen their synapse (or form a new synapse) when coactivated by the co-occurrence of the corresponding stimuli. However, repeatedly holding the spoon without hearing a buzz and, vice versa,

The Capability of Acquiring New Knowledge

hearing the buzz without holding a spoon would progressively weaken and eventually disconnect those "spurious" synapses. In contrast, the more frequent association of buzzing beetles would result in a robust connection between the respective neural representations. This view is broadly accepted and constitutes a tenable explanation for knowledge acquired by multiple repetitions and accumulated through extended experience.

However, a mechanism requiring repeated reinforcement to form memories does not explain the most familiar cases of one-trial learning of new facts. The day after hearing the buzzing beetle while eating ice cream we indeed know that beetles can buzz. In contrast, even though we have not handled a spoon or heard a buzz after that last encounter, we do not hold the incorrect belief that spoons can buzz.

The above issue is a serious challenge to neural theories of learning because every time we experience a "real" association, we are also simultaneously exposed to a much greater number of random co-occurring stimuli. The incorrect storage of such spurious information would systematically overwhelm the single piece of worthy knowledge acquired in the process.

What does this have to do with axonal-dendritic overlaps? It turns out that the third principle provides an exceptionally fitting protection against this cognitively catastrophic indiscriminate learning. That beetles could buzz is consistent with previously stored background information: flies, wasps, and bees buzz, fly erratically, and crawl, and beetles fly erratically and crawl. In contrast, the notion of buzzing spoons is nowhere near the same level of plausibility: forks and knives don't buzz (let alone fly erratically, at least in most households)! In neural terms, the axons and dendrites of the neurons encoding for beetles and buzz overlap and can therefore form a synapse. In contrast, the axons and dendrites of the neurons representing the elements of the other spurious associations (spoon and buzz, ice cream and buzz, etc.) do not.

The ability of axonal-dendritic overlap and background information gating to filter spurious associations is computationally powerful. In fact some major commercial search engines have recently reinvented simplified versions of this capability, probably without realizing they were mimicking the design selected by evolution for our brains. Think, for example, of online shopping carts prompting you to consider just before checking out "customers who purchased this item also bought that one." This strategy is usually more effective than randomly advertising (typically unrelated) products. For example, if you bought two products for pet care, this system will suggest additional articles purchased by other likely pet owners. In contrast, customers purchasing infant formula might be urged to consider baby clothes. It would be pointless to offer baby clothes to someone ordering dog shampoo.

Intriguingly, the basic tree-shaped structural elements of the nervous system (axons and dendrites) implement that same filtering function hundreds of billions of times every instant throughout each brain by carefully guiding synapse formation. Rather than "spamming" all neurons with the chance to synapse with irrelevant partners, axons and dendrites judiciously select potential synapses based on which other neurons every neuron is already contacting. This is achieved simply by the physical requirement that new synaptic contacts can only be chosen among preexisting axonal-dendritic overlaps (figure 6.4).

Let's begin to distill this very central discussion in the entire exposition of this book. The brain is a mighty machine made of a huge number of neurons interconnected by an even more humongous number of chemical synapses. Neurons transmit and receive electrical signals through spectacularly elaborated tree-shaped antennas: axons and dendrites. Axons are structurally more extensive and are responsible for much of the circuitry. Dendrites are biophysically more sophisticated and take care of most of the computation. This powerful network communicates with its physical surroundings by activating muscles and glands as well as by monitoring input from a plethora of sensory receptors.

The most dramatic, mysterious, and arguably most meaningful occupations of the human brain, and by inference of the nervous systems of many other animals, are internal. These functions are spelled out in an escalation of functions that we have summarized in three principles of the brain-mind relationship. First, the constant exchange of electrical activity yields all mental states we experience. Second, the continuous tweaking of the synaptic connectivity among neurons alters the knowledge of each individual (that is, the available repertoire and specific content of mental states), in every single moment throughout the life span. Third, the instantaneous patrimony of neuronal connections strongly constrains the selection of synapses an individual can form anew.

In this fantastic organization that is fundamentally determined by axonal and dendritic arborization, thinking, knowing, and learning are

The Capability of Acquiring New Knowledge

Figure 6.4
Convergence toward an idea (Daniel Segrè, 2014, Fresh Pond Park, Cambridge, Massachusetts).

Figure 6.5
The structure-activity-plasticity relationship of the brain (and mind). (Top) The instantiation of specific firing patterns in cell assemblies (corresponding to mental states) requires adequate circuitry (corresponding to knowledge) and triggers synaptic formation and pruning (corresponding to learning). These plastic changes obviously alter the structural connectivity of the network but are at the same time (and somewhat less trivially) also constrained by it due to the requirement of axonal-dendritic overlap. Mental states are intended here as subjective experience, which is traditionally viewed as a consequence of external reality but on purely logical grounds might well be the source of that external reality. (Bottom) The above mechanisms are enabled by the dense overlaps of neuronal arbors, such as those of this layer 2/3 chandelier cell from the rat motor cortex (see note 21 in chapter 3). The soma and dendrites are colored white, the axon brown. The two left insets zoom in at low and high magnification (top and bottom, respectively) on the arbors near the soma; the two right insets zoom in on more distal axonal regions (rendering by Amina Zafar in the author's lab). The reconstruction is freely available online at NeuroMorpho.org (branch thickness was increased to enhance contrast).

inextricably intertwined (figure 6.5). Thinking necessarily depends on what is known, and what is known is a result of what is learned. At the same time, learning is deeply contingent on both the present experience ("thinking") and background knowledge (past "learning"). Rather than constituting a computational burden delimiting the mental capacity of the brain, this interlinked gating relation between prior and future experience is an adaptive evolutionary trait both at the population and the individual levels.

Most poignantly, the prerequisite of axonal-dendritic overlap for synaptic alteration provides the unique cognitive advantages of avoiding the incorrect storage of an enormous amount of mistaken information corresponding to random co-occurrences of unrelated aspects throughout our bustling inner lives. It is sensible to say that it is indeed useful to learn less in order to know more. Requiring an appropriate level of prior knowledge ("leave it to the experts!") for learning more is truly a blessing rather than a curse.

6.4 Arbor Plasticity and Learning: Spatial and Temporal Scales

We are ready now to reassess from a more mature perspective an important puzzle we considered at the beginning of this chapter. If prior expertise

The Capability of Acquiring New Knowledge

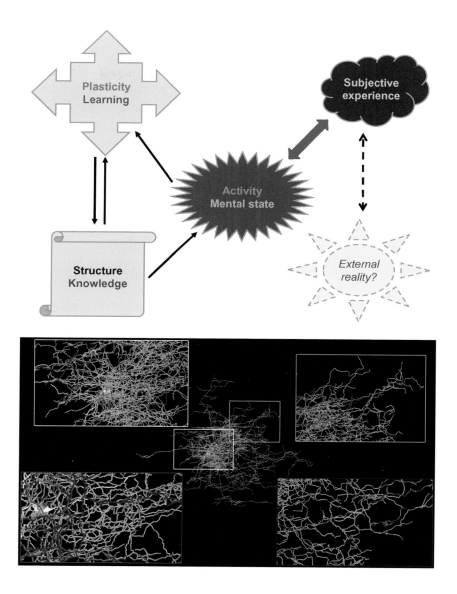

gates learning, how do we acquire completely new skills and knowledge? How can we decide one day that it's time to learn bridge, pick up a beginner's book in the library, take a couple of introductory lessons at the nearby social club, try out a few times with benevolent friends, and become decent players within the year, even though we have no previous experience playing cards? In section 6.2 we planted the seeds of the answer by duly noting that background information only gates one-trial learning. Although much of what we know is indeed the fruit of single experiences, such as a serendipitous realization, a conversation with a teacher, reading an inspiring book, watching a movie, or observing a buzzing beetle, learning to play bridge from scratch requires long days of dedicated effort. Perseverance and rehearsals are a trusted recipe for transitioning from novice to masterful performance, be it behavioral (a golf swing, a cello recital, or a public speech) or cognitive (from preparing for a school exam to learning a foreign language).

How can persevering repetition circumvent the apparently inescapable need of axonal-dendritic overlaps for synaptic formation? This follows a touch of (apparent) magic: axons and dendrites resemble the trees of an enchanted forest rather than the ordinary arbors in your backyard (figure 6.6). If you stare at a neuron long enough, you will see slow and rare movements of its axonal and dendritic branches. Indeed, although the vast majority of the neuronal arborization in an adult brain is fairly static, at any one time a few of the trees wiggle their limbs by tiny bits. Some branches retract, others extend, and yet others bend or shift in space. Over time, these small movements can amount to a comprehensive reorganization of the network, not just as far as connectivity is concerned but most consequentially in terms of axonal-dendritic overlaps. As a result of these neuronal rearrangements, whereas in one given moment the ability to learn is in fact gated by prior knowledge, sufficient training and/or consistent study possess the ability to remold the very blueprint of the learning machinery!

How rare, slow, and tiny are these arbor alterations? Can they really account for one's ability to pick up a complicated game such as bridge without any previous experience in playing cards? The biochemical machinery underlying these events is actually similar to the mechanisms at work when spines twitch in and out of dendrites or varicosities crawl up and down the axons to form and remove synapses. To understand how difficult it may be

Figure 6.6
Enchanted forests. (Top) Trees talking to each other (Daniel Segrè, 2014, Fresh Pond Park, Cambridge, Massachusetts). (Bottom) Arborization (Daniel Segrè, 2014, Halls Pond Sanctuary, Brookline, Massachusetts).

to move an entire axonal or dendritic arbor, imagine a botanical tree. Try to visualize chipping away minute pieces of bark the size of unopened flower buds from one branch and progressively piling them up on another branch. The analogous process in the mammalian brain consists of pruning or accumulation of cellular membrane along the branches of axons and dendrites. This movement has been recently monitored in the mammalian brain. In rodents, only a small fraction (between 3% and 20%) of the axons and dendrites undergo this process at a given time, and the pace of remodeling amounts to a minuscule one thousandth of 1% per week.

If this rate seems frustratingly low, consider for comparison the stability of our mind. Remember, however, that these rearrangements of entire axonal and dendritic branches correspond not to the agile learning of quick facts ("beetles can buzz") but to profound changes of our personality ("I'm a bridge player"). These more radical events are rather exceptional in the course of one's lifetime, and they typically only involve certain cognitive or behavioral aspects and not others. For example, the skills necessary to learn bridge are rather different from those required to train as an airplane pilot or to become an expert photographer (in all cases starting from complete naïveté). In truth the skill set of most people remains depressingly static over the course of a typical month.

To recap the above considerations, both major arbor rearrangements and the corresponding adaptation of the brain learning aptitude derived by renewed axonal-dendritic overlaps constitute an ever-present, if very sluggish, flux more akin to the movement of glaciers than to the flow of a river. These nearly undetectable transformations explain the brain's uncanny capacity to adapt to new environments and conditions.

To extend the illustrative scenarios depicted at the end of section 6.2, suppose that your cousin visiting from out of town, who was present when your colleague first gave you directions to her house, ends up dating your colleague, and returns to see her every weekend. Within a few months of driving in the area, your cousin learns the way in and out of the neighborhood even better than you. The axonal and dendritic branches in his brain have rearranged (in light of repeated experience) to allow synapse formation encoding for a mental map and effective route finding.

Although we have focused so far on adult brains, the nervous system is far more plastic during development. Axons and dendrites wave their branches more often and more expansively in young brains, and the mind displays a correspondingly greater flexibility in its youth to seamlessly acquire new expertise. The superior neuronal plasticity of the younger brain explains the greater learning ability of children relative to adults. The next chapter expands a little more on the differences between newborn and aged neural trees. Without a discussion of the details of migration and differentiation, a brief preview of neural development is useful to provide a more complete picture of brain plasticity. Although most neurons are generated in the embryo, new neurons still continue to be added in early fetal stages. Neurons progressively stop being incorporated into various regions of the brain at different time points during development, both before and right after birth. Interestingly, the only region of the human brain that keeps producing neurons throughout adulthood is that most crucially responsible for learning and memory: the hippocampus (also mentioned in section 1.2). We revisit the differences among brain regions and their neurons in the next chapter.

We have discussed a broad range of temporal and spatial scales of activity, plasticity, and function, and table 6.1 summarizes the main elements of this multifaceted construction. At the fastest extreme, neurons fire spikes each lasting approximately one-thousandth of a second. Spikes are mediated by movements of tiny ions whose diameter is smaller than one-hundredth of the thickness of the smallest neuronal branch. A pattern of dozens of spikes in each of thousands of neurons constitutes the quickest mental states we can experience, of the order of a tenth of a second. These

Table 6.1
Mechanisms and temporal scales of brain-mind plasticity

Brain	Mind	Temporal Scale
Circuit spiking patterns	Mental state	0.3–30 seconds
Synaptic strength change	Thought likelihood	0.3–30 seconds
Synaptic formation/pruning	Knowledge acquisition	3 seconds to 30 minutes
Axonal-dendritic overlap	Immediate learning ability	3 minutes to 3 hours
Axonal/dendritic growth/retraction	Learning by training	3 days to 3 years
Neurogenesis and neural migration	Development of new capacity	Entire lifespan

electric activity patterns propagate from one neuron to another by chemical signals across synaptic connections.

Such lines of communication allow circuits to chain together consecutive waves of activity representing longer-lasting mental states, typically up to several seconds or even minutes. In addition to forming the neural correlates of mental experience, extended firing sequences also alter the strength of the underlying synapses. These modifications take only seconds to occur but can last for an entire lifetime. They are mediated by complex molecular machineries that are much (more than ten times) larger than the ions underlying spikes but still much smaller (also more than tenfold) than the thickness of the smallest neuronal branch.[11] The mental correlates of these events are changes of how prone we are to experiencing the mental content implicating those synapses.

The endless flow of activity in the nervous system also reestablishes the very presence and location of connections among neurons, creating some new synapses on appropriate axonal-dendritic overlaps and eliminating others. These crucial connectivity touch-ups may take from a few seconds to several dozen minutes and reflect the quintessential process of learning or acquiring the capacity to experience a new mental state. The key players in such a game of structural plasticity are dendritic spines and axonal varicosities, whose size is comparable to the thickness of neuronal branches. Last, at the slowest end of the scale, new neurons can appear, and existing neurons can rearrange their branches during the course of several months and over the entire arbor, invoking a spatial span thousands of times larger than the branch thickness. These changes correspond to slow alterations of expertise and require persistent repetition. By modifying the compendium of axonal-dendritic overlaps, these arbor rearrangements fundamentally affect the capacity to form new synapses and thus the ability to learn further information.[12]

Before closing this very central chapter, we should remember once more that the neural correlates of mental states are cell assembly activity patterns rather than spikes in individual neurons. The last chapter distinguished between knowledge as the *capability* of instantiating mental states (conceptually corresponding to synaptic connectivity) and the probability of eliciting a given mental state (determined by synaptic strengths in the circuit). If mental states were represented by single neurons, the above correspondences of synaptic connectivity as knowledge and of synaptic strength as probability of experience would be strictly literal. Instead, in the last chapter we showed that coding by cell assemblies somehow blurs the line between these two aspects. As a result, synaptic strength occasionally affects the capability of instantiating a mental state, and synaptic formation similarly influences the probability of experience.

Coding by cell assemblies could have the same effect on the correspondence between axonal-dendritic proximity and the ability to learn. In other words in many circumstances it may be fair to say that axonal and dendritic branching determines the *probability*, rather than *possibility*, of learning. To understand this, consider again the example of the first experience of a buzzing beetle. Suppose that the sight of the beetle is encoded by the coactivation of any eighty of one hundred "beetle" neurons, and, similarly, that the buzz sound is encoded by the coactivation of any eighty of one hundred "buzz" neurons.[13] To learn that beetles can buzz, we need to form at least eighty new synapses. There are a total of 10,000 pairings between the dendritic arbors of the beetle neurons and the axonal arbors of the buzz neurons (100 times 100). At a very minimum, learning this association requires axonal-dendritic overlaps in eighty of those 10,000 pairings. If exactly those eighty beetle neurons and those eighty buzz neurons that happened to overlap get coactivated to encode the "buzzing beetle" experience, it is possible that eighty new synapses would be formed. Then the next time we hear a buzzing sound, if that experience is encoded by exactly the same eighty neurons, we might have the thought "it could be a beetle."

Learning that beetles can buzz in the above scenario of only eighty axonal-dendritic overlaps, however, would be unlikely, and the learned association would be quite weak. The reason is that there are many combinations of eighty beetle neurons and eighty buzz neurons that could encode for the buzzing beetle experience, and the chance to match exactly the right 160 cells is very small. Moreover, even if the learning jackpot were hit and these eighty synapses were formed, the likelihood that any subsequent buzzing experience would be encoded by the same eighty neurons (thus eliciting the thought of a beetle) would also be low. Now consider the opposite scenario, in which axonal-dendritic overlaps are present for each and every one of the 10,000 pairings between beetle neurons and buzz neurons. In this case, any possible neural correlate of the buzzing beetle experience would result in sufficient synaptic formation to ensure strong association between the buzzing sound and the thought of beetles.

The two opposing scenarios of sparse versus dense axonal-dendritic overlaps we have depicted represent the extremes of a broad spectrum of possibilities. Typically, the number of pairings between beetle and buzz neurons in which an axonal-dendritic overlap is present will fall somewhere between 80 and 10,000. If this number is low and close to eighty, the chance to learn will be small and the resulting association weak. If the number is high and close to 10,000, the chance to learn will be big and the resulting association strong. Thus, although in the simplified grandmother neuron view of the brain, axonal-dendritic overlap indeed corresponds to the binary (yes/no) capability of learning, in the more general and realistic model of cell assemblies both the probability and strength of learning increase with the number of axonal-dendritic overlaps.

Part III: A Bucolic View of the Universe

7 Neurobotanical Gardens

7.1 Worms, Flies, and the Rest of Us

The three principles of the brain-mind relationship we have now explained are based on fundamental structural and functional features that are common to the majority of neuron types in most brain regions of nearly all mammals through every stage of the life span. One of the most mesmerizing aspects of neurons, however, is the stunning diversity of their shape and of other biophysical and molecular characteristics. Having understood how the tree-like shape of axons and dendrites *in general* relates to neuronal function and corresponding phenomena of the mind, we can engage in a tour of neuronal diversity with a deeper appreciation of its implications. Because axons branch out to connect to other neurons, and sophisticated computational operations take place within dendritic arbors, the differences in these trees among neurons have deep functional consequences on the possible firing patterns and thus mental states. Furthermore, because axonal-dendritic overlaps correspond to the ability to acquire new knowledge, the variety of shapes we are about to encounter has direct implications for the type of learning that various neurons can be involved in.

Each individual neuron is different from every other one (much as any two plants in a garden), as no two living objects are ever exactly alike in nature. Even a given neuron does not remain the same over time, as we have seen in the last chapters, because of its continuous structural and biophysical adaptation to activity and experience. Over time the cumulative changes over the life span make the neurons of a young brain quite different from those of the same, aged, brain. Because different individuals live different lives, even if they started with exactly the same neurons,[1] they would rapidly diverge into distinct shapes and networks. Newborn brains already start as nonidentical collections of neurons, so adult nervous systems capture an even greater diversity of properties.

More strikingly, even a single contained region of one brain is home to a great variety of diverse shapes (akin to different tree types, such as pine and oak). These different neuron types reflect distinct computational functions within the network. The diversity of neuron types becomes all the more extreme when we consider various parts of the brains, not dissimilar to the different vegetation that can be found in different continents and climates. Such major differences also relate to the specific functions carried out by each brain region.

We consider all of the above sources of neuronal diversity in this chapter and the next. For now, let's start with perhaps the most important distinction to differentiate neurons, namely the broad variety of animal species possessing a nervous system. The only reality we know, as human beings, is that produced by the neurons of our own nervous systems. Humans, however, are just one kind of primate, and primates are only one kind of mammal (figure 7.1). Studying the neurons of the human brain presents obvious difficulties. The experimentation that can be carried out on excised brains post mortem is quite limited because too little of the function remains after the death of the organ. Neuronal shape can still be examined in theory, but in practice even the structure of neuronal trees, and especially axons, remains as of today nearly impossible to probe in the human brain.[2]

Alternatively, neurons can be studied on small pieces of brain tissue removed during surgical procedures, for example when ablating brain tumors. However, these portions are obviously kept as small as possible and are thus too little to contain the exuberant expanse of a full axonal tree. Instead, we owe much of the knowledge about the neurons of the human brain to research on other mammals, especially monkeys and rodents. Monkeys are particularly useful because of their neurological and cognitive similarity to humans,[3] and most of our understanding of vision, the foremost human sensation, is derived from recordings in these creatures. For the majority of knowledge on the structure of dendrites and almost the entirety of that of axons, our debt of gratitude goes to rats. Mice have provided most of the recent data thanks to their superior genetic accessibility.

Human neurons tend to be bigger than those of rats and mice because of the greater dimension of the brain and also the larger number of synapses

Neurobotanical Gardens

Figure 7.1
Hidden primate (Daniel Segrè, 2014, Idaho).

Figure 7.2
Tiny or huge, squishy or crunchy, all multicellular animals (except sponges) have tree-shaped nerve cells. (Clockwise from top left) (1) A mechanosensory neuron from the somatic lumbar ganglion of the worm *C. elegans*.[4] The main trunk of this neuron (colored green) spans the length of the worm's body (approximately 1 mm) and branches off nearly identical subtrees (colored red and orange) on both sides at regular space intervals. The axon (yellow) breaks the symmetry by selectively innervating one quadrant (rendering by Uzma Javed in the author's lab). (2) A neurogliaform interneuron from layer 2/3 of the elephant visual cortex.[5] The horizontal span of this neuron is 0.8 mm. Soma and dendrites are colored green with blue bifurcations (rendering by Namra Ansari in the author's lab); the axonal tree was not reconstructed. (3) A nerve cell of the same type (neurogliaform interneuron) from the rat cortex[6] spans only 0.15 mm (enlarged twofold here relative to the scale of the elephant neurons for better visualization). Soma is colored purple, dendrites light blue with dark blue terminals (rendering by Uzma Javed in the author's lab); the axonal tree was not reconstructed. (4) A stomatogastric ganglion cell from the spiny lobster.[7] The soma and axon are colored brown; the dendritic tree, peculiarly stemming out of the axon, is in shades of green and blue (rendering by Namra Ansari in the author's lab). (5) A ganglion cell from the rat retina.[8] The soma and half of the dendritic arborization are colored red; the rest of the dendrites are blue, the beginning of the axon is gray (rendering by Amina Zafar in the author's lab). The rest of the axon, traveling from the back of the eye to the center of the brain, was not reconstructed. All reconstructions are freely available online at NeuroMorpho.org (branch thickness was increased to enhance contrast).

they tend to make. Interesting comparisons have also recently been drawn with the neurons of even larger mammals, such as elephants and giraffes (figure 7.2). The axons of certain neurons in the motor cortex of the giraffe, for example, reach from the top of the brain down to the end of the spinal cord near the tail. These axons are more than 20 feet long and still only less than a thousandth of a millimeter thin! Unfortunately, the sheer size of these arbors also constitutes a formidable technological challenge for their scientific characterization. To this day our knowledge of neurons of any large animal is mainly limited to the dendrites.

Still, all mammals together are but a tiny proportion of the set of creatures with nervous systems. The brains of certain birds, for example, are subject to intense research because of their capacity to learn complex songs, constituting a useful model of sequential learning. The nervous systems of some amphibians (such as frogs) possess a remarkable ability to adapt to drastic changes. The neural circuits of some frogs, for instance, restore their

Neurobotanical Gardens

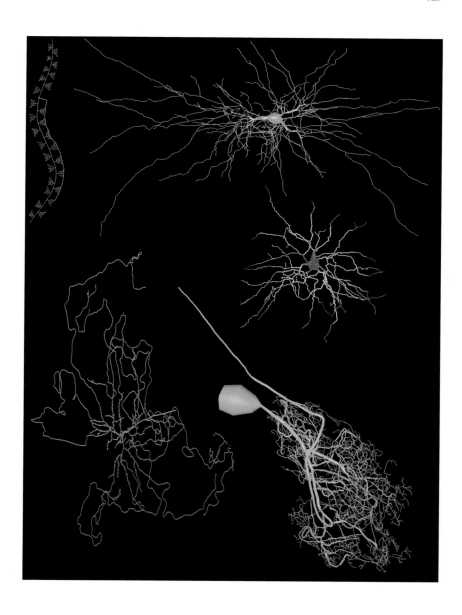

vision after their eyes are surgically flipped upside down! Such capacity offers a rare opportunity to investigate the mechanisms underlying structural plasticity. There are also several species of transparent fish that offer unique windows into the structure and activity[9] of neurons, both figuratively and literally, because their brains can be examined directly under a microscope.

All of the above examples, however, happen to involve vertebrates, which are still just a minority of the animals endowed with neurons. As we move farther from humans in taxonomic and genetic similarity, we encounter an even broader variety of creatures, including insects, mollusks, crustaceans, and other boneless animals. The neural organization of these species is fundamentally different from that of humans, whose brain and spinal cord (together constituting the *central nervous system*) are encased in the skull and vertebral column, respectively. In certain ways invertebrates may be a better model for the human *peripheral nervous system*. The peripheral nervous system consists of neurons connecting to skin receptors and to internal organs, such as the kidney or liver. These neurons, unlike those in the brain, are not encased in bones. Most importantly, the nervous systems of invertebrates are relatively simpler than those of mammals, thus offering a glimpse of hope for a fuller functional understanding. The large size of the axons of certain neurons in the squid, for instance, enabled the historical recordings of electric activity that led to the discovery of action potentials.

Two of the most intensively investigated species in neuroscience—aside from rats and mice, which we discuss more extensively in the rest of this chapter and in the next one—are a humble worm called *Caenorhabditis elegans* (*C. elegans* for short) and the common fruit fly. *C. elegans* is the known organism with the fewest number of neurons, just 302 per animal.[10] Nonetheless, with such limited machinery, the worm manages to crawl around with an elegant wavy motion (hence the name), to find and ingest nutrients, to recognize suitable mates and reproduce, and to avoid noxious conditions such as toxic chemicals or excessive temperatures. Furthermore, individual *C. elegans* worms are very similar to one another with respect to their nervous system. Specifically, the same neurons can be found in the same positions and nearly matching shapes and connectivity from animal to animal.[11] Moreover, given the limited number of neurons in *C. elegans*, every nerve cell forms a "neuron type" by itself, likely performing a distinct function as if it were a "grandmother cell."

On the one hand such precisely identifiable replicas across animals should facilitate the investigation of the structure-activity-function relationship. On the other, this same repeated and deterministic organization also raises serious doubts of relevance to the mammalian (and particularly human) brain. These much more complex nervous systems, as we have explained, display, in stark contrast to that of the worm, extreme diversity and code information by way of cell assemblies each formed by hundreds or even thousands of neurons from dozens of different types. Furthermore and most importantly, unlike in more complex nervous systems, the same trees serve both as axon and dendrite in the neurons of *C. elegans*, with electric activity flowing both to and from the cell body.

Despite the simplicity of this worm nervous system, we still do not know how the neural network of *C. elegans* works! This is sometimes taken as evidence that knowing the full connectivity of a neural circuit may not suffice to infer its computational function. Unfortunately, very little electrical recording has been performed to date in *C. elegans*. Perhaps, if we knew as much biophysics about the neurons of *C. elegans* as we do of the 300 best characterized neurons in the rat, we could indeed write a satisfactory "engineer's handbook" of the worm. Most research on this animal until now has focused on cellular structure and molecular genetics.

The fruit fly (technically called *Drosophila melanogaster* or simply drosophila) has recently emerged as a leading animal model of neuroscience. Its fast reproduction cycle and relative genetic accessibility have allowed the creation of a large number of genetic mutants. These include strains in which specific neurons are conveniently color coded for easy recognition under a microscope or can be switched on and off with laser beams. The nervous system of drosophila is not as simple as that of a worm but is still more tractable than that of humans. It has a couple of hundred thousand neurons, an as-yet-undetermined proportion of which can be reliably recognized in each and every individual. Drosophila routinely engages in very sophisticated behavioral and cognitive tasks, including learning and memory, flight, navigation, social interactions, and purposeful fighting. Yet some aspects of its biology distance this animal from humans. For instance, drosophila undergoes a metamorphosis during development, passing through several larval stages after embryonic maturation and before fly adulthood. From the perspective of neural organization, the larva appears to be an almost entirely different animal (no pun intended!). At least the initial organization of the adult fly is likely to be more under deterministic molecular control than even the newborn human brain.

Given the considerable differences between drosophila, and especially *C. elegans*, and the targets of our interest, the human brain and mind, it is highly questionable whether these species constitute suitable animal models with respect to the three principles presented in chapters 4, 5, and 6. A healthy dose of skepticism is particularly appropriate when considering the complex interactions between experience and learning, network structure and knowledge, activity patterns and mental states. We thus swiftly return to the mammalian brain in order to continue our exploration of neuronal diversity, but with a word of caution: mammals are newbies in evolutionary terms. It's useful (and humbling) to remember that many more species exist with very different nervous systems from ours. These species functioned in a continuously changing environment for millions of years and still thrive today.

Let's thus keep *C. elegans* and drosophila in mind if only as mementos of two nontrivial facts. First, even in such distant evolutionary cousins of ours, neurons display some of those same fundamental architectural elements that remain conserved in all species, namely tree-like branching cables subserving electric conduction and network connectivity, appropriately integrated to result in functional circuits. Second, when assessed from a genetic and evolutionary perspective, rats and humans have eerily similar brains, at least at the level of their neuron types and properties. Although many aspects of network organization manifest in strikingly similar ways in both rodents and primates, some dissimilarity is also evident. The prominent behavioral and cognitive differences that distinguish humans from other animals are most likely attributable to the relative and absolute size of the neocortex, resulting in qualitatively and quantitatively superior computational capacities. We return to this consideration in chapter 9.

7.2 Arbors for All Seasons

How does the mighty human brain come to be? The development of the central nervous system during gestation is nothing short of majestic.[12] Along with the appearance of other organs during embryonic growth, the brain begins to form from the neural tube around the fourth week after fertilization. At this early stage the brain "seed" resembles an unimpressive hollow sac, but this deceptive appearance hides biochemical machinery so complex and finely coordinated as to resemble the busiest, largest, and technologically most advanced of imaginable factories. In an empty-looking fold toward the middle of this formation, neural progenitor cells keep splitting and differentiating. Initially, these neural stem cells simply divide like many other cells during growth, each forming two nearly identical progenitors. This process creates an accumulation of progenitor cells and a corresponding enlargement of the "protobrain."

At the beginning of week 8, these cells progressively switch to a different mode of *asymmetric* division, each generating one progenitor cell (thus maintaining the generative capital) and an undifferentiated neuron. By the end of week 9, only a minority of neural progenitors split by symmetric division. At this point the nervous system is already much more complex than the simple neural tube it had started as. During this same six-week period, the embryo has grown by a factor between six and ten. Asymmetric division of nervous system cells, however, continues for at least another seven weeks. Throughout this whole period, the undifferentiated neurons begin to migrate while starting to extend branching structures.

When neurons cross the target region of their intended axonal connections on their way to the expected destination of their soma and dendritic tree, they launch an "anchor" in passing. This extension remains in the target region and starts sprouting a tree while the neuron is still migrating (progressively stretching the arborization). In contrast when neurons reach their destination without entering the target region of their axons, they begin elongating both axonal and dendritic trees shortly thereafter. Such an orchestration advances like a marching band of 100 billion instruments. As axons fill space, dendrites reach out to form overlaps and connections. In the short span of six months, more than a trillion synapses and 10 trillion axonal-dendritic overlaps bud out, a rate of approximately 100 million synapses and 1 billion axonal-dendritic overlaps *per second*!

The result of this grandiose neurodevelopmental "spring" is the most exuberantly hyperconnected network that brain will ever be. The newborn central nervous system contains its maximum number of neurons, of synapses, and of axonal-dendritic overlaps. At birth (or shortly thereafter), neurons start to die off, and their generation stops in almost every brain region. Some types of neurons are very abundant early in development but nearly disappear as the brain matures into adulthood. In all neurons synapses get pruned after birth at a faster rate than they form, and axons and dendrites on average retract,[13] resulting in a net reduction of axonal-dendritic overlaps.

Yet synaptic connectivity at birth had hardly any chance to be sculpted by experience. It thus serves as a mere placeholder in contrast to the wiring of the more mature circuit that effectively codes for knowledge. Much more consequential is the large number of axonal-dendritic overlaps, endowing the neonatal brain with a terrific ability to learn from experience by properly establishing synapses in an activity-dependent fashion. As a consequence, an infant brain knows nothing but can learn (nearly) everything.

The next phase of human brain development effectively lasts nearly thirty years, namely, for the entire childhood, throughout adolescence, and for a considerable period of what is socially considered young adulthood. This is perhaps the most intriguing and, as of yet, least understood season of the brain, forging and molding every part of this uniquely dynamic computational machinery like a scorching summer. The constant flux of the human brain during its first three decades is far more dramatic than the subtle adjustments we described in the last chapter. Whereas radically new knowledge can be acquired later in life only at the cost of considerable effort, and not always in optimal ways, learning novel information is easier, faster, and more effective in the very young brain. Children exposed to two languages, for example, grow up to be perfectly bilingual adults, whereas anyone learning a second language later in life can be recognized for at least a slight accent or sometimes peculiar word choices. Moreover, children recover often completely from massive strokes or the surgical removal of large portions of the brain, whereas these same events are typically devastating in adult individuals.

Whether the superior plasticity of young brains is due to much denser and more extensive axonal-dendritic overlaps, to greater branch mobility of axons and/or dendrites, to a combination of both, or to other mechanisms altogether is not yet firmly established.[14] What is already clear, however, is that this wave of "summer storms" does not hit the entire brain at once. Certain more basic computational processes, such as right-left eye coordination, are plastic very early and get consolidated within a few years of life. More sophisticated cognitive skills, such as the ability to attribute an independent mind to other people, develop a few years later and remain plastic for longer. The exact mechanisms underlying these critical windows remain only partially understood.

The staggered and inconsistent development of different brain functions parallels the complexity of the whole body. The age at which children become adult from the point of view of a dentist may differ from that of a cardiologist or endocrinologist. Similarly, the legal notion of adulthood depends on the context (e.g., schooling, driving, smoking) and on local jurisdiction. In the brain different functional regions remain "children" for different times. Whereas the visual and motor cortices are certainly adult in a twenty-year-old person, the prefrontal cortex is still in its early adolescence at that time. The determination of when the human brain can be said to be mature is rendered even more complicated by the difficulty of mapping development across different species. Much of what we know about neurons is based on rats and mice; it is challenging if not impossible to draw a precise correspondence between the relative maturation of different regions in the rodent and human brain.[15]

With the eventual maturation of the latest regions, the human brain enters its most stable period in the late twenties. The ensuing adult phase compares to the fall season. Even then, certain areas of the brain maintain a defined level of structural plasticity. This residual plasticity ensures the ability to form new memories. Specifically, adult plasticity in different brain regions relates to the storage and processing of distinct types of memories.

Semantic memories, the knowledge of facts (e.g., the state capitals), mostly involve the neocortex. Autobiographic and prospective memories, the ability to recall past episodes and to formulate future intentions in one's life, leverage the activation of the hippocampus. Emotional memories, such as fear responses, employ the amygdala. Procedural memories and habits, behaviors that are typically performed on "autopilot" (such as the mechanical sequence of manual handling to prepare coffee first thing in the morning), require the basal ganglia. Learning precise coordination, most readily observed in fine motor control (such as trying a new piano piece) but also likely occurring in "mental gimmicks," recruit the cerebellum.

The main considerations we have developed in the past two chapters are particularly relevant in these important regions. In this period of "conditional plasticity," synaptic turnover is more stringently dependent on axonal-dendritic overlaps. Personalities are now well formed, and expert knowledge is powerfully harvested from daily experience.

But after fall comes winter. With aging, the trees of the brain start losing their synaptic "leaves." Dendritic spines and axonal varicosities become sparser, reducing the number of connections. The continuous reduction of neurons due to the faster rate of neuronal death relative to regeneration accelerates, as neurons die faster and regenerate even slower. All in all, a substantial proportion of the adult network is affected by these changes, and approximately 20% of connectivity is lost. This transformation weakens the circuit, making it occasionally more difficult to retrieve memories and slowing mental reflexes. "Successful" aging, however, leaves the cognitive and computational functionality of the elder brain still remarkably intact. This is no small achievement: after all, the second principle equates knowledge to circuitry, yet healthy retirees do not forget one in five of their memories. In comparison, randomly cutting 20% of the wires in a computer would likely result in its catastrophic failure.

Although once again the mechanism underlying such remarkable stability of healthy aging brains is not known, we can at least speculate as to the most likely hypothesis. Throughout life, the nervous system continuously morphs its circuits in response to experience. It does this countless times over thanks to its massively parallel constitution. The hundred billion neurons, thousand trillion synapses, and ten thousand trillion axonal-dendritic overlaps can encode, we propose, even more mental states than necessary to represent the entire cognitive content of an individual. Thus, the network automatically stores every memory several times with redundant and independent cell assemblies. The result is a high degree of built-in robustness. If a neuron fails, if a synapse disappears, if an entire cell assembly flounders, an equivalent firing pattern can still be instantiated via alternative pathways, a tad slower and less rehearsed perhaps but still doing the job.

Eventually, however, the network thins down to its last wires. How many years it takes to reach such a brittle state varies from individual to individual, and needless to say we do not yet know the exact interaction of genetics and lifestyle that determines this turning point.[16] The end game is characterized by an unavoidably steep cognitive decline, as memories and skills severely deteriorate. The lone residual pathways in which they were encoded fall victim to synaptic loss and neuronal death. This catastrophic deterioration is dramatically accelerated by pathologies such as Alzheimer's and Parkinson's diseases. The former causes neurodegeneration in the neocortex and the hippocampus, resulting in a progressive and rapid loss of autobiographic and prospective memories and eventually semantic knowledge. The latter targets the basal ganglia, damaging procedural memories and habitual behaviors and leaving in their place the dreadful signature of generalized tremors.

Much current research focuses on how to slow down or even arrest physiological and pathological cognitive aging. For example, compelling data demonstrate that, at least in certain regions of the brain such as the hippocampus, the dendrites of some of the main neuron types become bushier and bushier with aging (figure 7.3). One possible explanation is that when many of the axons that used to be presynaptic to those neurons die off, the dendritic arbors attempt to react by sprouting more branches. The resulting higher dendritic density could thus represent the receiving neuron's attempt to counterbalance the reduction in the number of synapses and of axonal-dendritic overlaps in the face of decreasing axonal innervation. Obviously, however, such compensation might profoundly alter the fine connectivity of the circuit and therefore its knowledge and activity patterns.

A much more comprehensive understanding of the relationship between the axonal-dendritic networks and computational function will be necessary to design interventions to prevent, cure, or circumvent neurodegeneration. The same knowledge will, we hope, enable every individual to make informed choices in the winters of their lives. We briefly discuss the most promising technologies and their deep philosophical implications at the end of the book.

7.3 Geographical Diversity and Physique du Role

Almost every chapter in this book refers to several distinct regions of the brain, each with its computational specialization. The building blocks of all brain regions are always neurons with their axonal and dendritic arbors. Do the shapes of axons and dendrites vary to reflect the functional differentiation of the networks they form? The answer to this simple question is "yes, but to different extents," revealing once more the pervasive neuronal diversity of the brain.

In some cases neurons display peculiarly idiosyncratic branching patterns that make them uniquely distinguishable as belonging to a specific

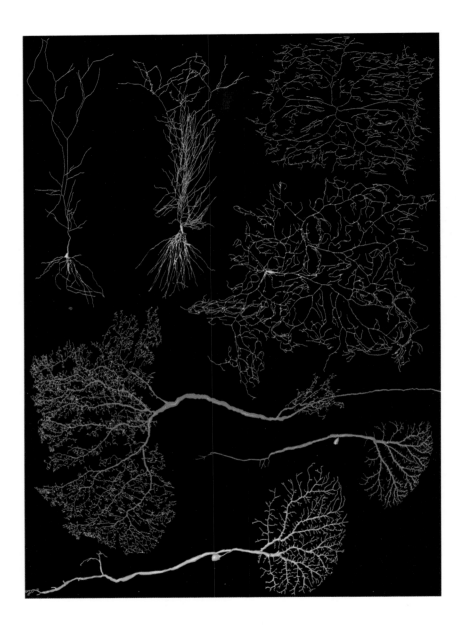

Figure 7.3
Changing brains in changing bodies. Clockwise from top left: (1) A CA1 pyramidal neuron from the hippocampus of a two-week-old ("pubescent") rodent.[17] The soma and basal dendrites are colored yellow (with darker terminals), the apical dendrites blue (with lighter terminals); the axon was not reconstructed. (2) A neuron of the same type (CA1 pyramidal) and from the same region (hippocampus), but in a two-year-old ("elderly") animal[18] (color codes as in the first neuron). The older neuron has a much richer arborization (over 400 branches vs. fewer than 100), but the overall shape is recognizably similar. (3) A sensory neuron from the fruit fly larva[19] (pre-metamorphosis) with main dendritic trunks colored purple, inner terminations green, and outer terminations blue. (4) A neuron of the same type, species, and developmental stage but different individual (see note 14 in chapter 3), with dendrites colored blue and bifurcations highlighted in yellow. (5–7) Three sensory neurons from the adult blowfly[20] (postmetamorphosis), with the main dendrites colored, respectively, purple, red, and blue and the terminals highlighted in yellow or green. The somata are visible only in the bottom two neurons (colored green and blue, respectively) (renderings by Uzma Javed [1, 2, 3, and 5], Namra Ansari [4 and 6], and Amina Zafar [7]). All reconstructions are freely available online at NeuroMorpho.org (branch thickness was increased to enhance contrast).

brain region. As an example, Purkinje cells, among the most recognizable neurons due to their characteristic dendritic tree, are exclusively found in the cerebellum (see figure 4.3 in chapter 4). In other cases the same general neuronal shape is found in several brain regions. For instance, pyramidal cells are the principal neurons in all the regions of the neocortex, in the main portions of the hippocampus, in the amygdala, and elsewhere. At the same time pyramidal cells from the neocortex, from the hippocampus, and from the amygdala are not exactly the same. In fact, expert neuroscientists readily recognize neocortical pyramidal cells even from different layers (that is, depths from the cortical surface) as well as from distinct subregions of the neocortex. Because pyramidal cells are present in five different layers and at least two dozen functional areas, it is possible to catalog more than one hundred distinct types of pyramidal cells in the neocortex alone!

The situation is yet again enchantingly analogous to that encountered for botanical trees. In every continent, characteristic vegetation can be readily identified with specific geographical locations. Palm trees are most typically found in hot climates near the ocean or sea, as in Florida, Southern California, Morocco, Micronesia, and the Caribbean. In contrast, pine trees thrive in cold, dry climates typically found on or near mountain ranges, such as on the Alps and in Canada, and cactuses live in rocky deserts in Patagonia, North America, and Sri Lanka. Each of these families of trees (like cortical pyramidal cells) counts a large number of distinct species, each of which is often largely found in a restricted subregion. Yet other tree types are commonly found almost everywhere in the world, including birch, basswood, oak, beech, alder, ash, and many others.

But the metaphor only holds so far because the structure of dendrites and axons ultimately reflects their computational function. Thus, the specialization and consistency of the shape of various neuron types across brain regions might be linked to corresponding functional differentiations and commonalities in different subcircuits of the nervous system.

Among the most distinguishable neural trees are those that allow interaction with the environment, namely those of sensory neurons and motor neurons. The most important sensation in humans is undoubtedly vision. Visual signals are first captured by photoreceptors, special neurons in a dedicated layer of the retina, which is lodged at the back of your eyes. Instead of dendritic trees receiving synaptic connections from axons of other neurons, photoreceptors extend unbranched protrusions responsible for detecting light by converting it into electric signals. There are two main types of photoreceptors, named after the respective shapes of their "input" extensions: rods and cones. Rods are specialized to respond to dim light, do not discriminate among colors, and are more numerous in the retinal periphery. In contrast, cones are specialized to respond to bright light of specific wavelengths (colors) and are more abundant toward the center of the retina.

Inside each rod is a stack of inner membranes called (and shaped like) disks, each containing molecules that undergo a structural rearrangement when hit by light. By changing their structure, these molecules lose affinity for and therefore detach from other (receptor) molecules, sort of like a bent key no longer fitting into its hole. A series of these reactions eventually closes an ionic channel, discontinuing the inflow of positively charged sodium into the rod. This results in a change of electric potential across the membrane, back to the usual way of information processing in neurons. The process is similar in cones, except that it occurs in folds of the neuronal membrane rather than in inner disks. Note that the transformation of light into electric signals involves similar (but opposite!) mechanisms as encountered in standard synaptic neurotransmission, described in section 3.2: in "normal" dendritic synapses, neurotransmitters bind to a receptor, opening an ionic channel, which results in sodium influx. In retinal photoreceptors, light causes "unbinding," channel closure, and arresting of sodium influx. From this perspective, one could say that rods and cones are sensitive to darkness rather than light!

In essence photoreceptors transform a different form of energy (light, that is, electromagnetic waves) into electricity, a process called *transduction*. Analogous processes occur in all other sensory modalities. Tactile sensations, for example, are mediated by mechanoreceptor neurons. These cells are located just outside of the spinal cord and extend two long processes. One of them, consonant with a classic axon, outputs action potentials to the rest of the nervous system. The other axon-looking extension reaches the monitored piece of skin (e.g., the tip of the right index finger) to relay inputs, thus performing the duty typical of dendritic trees (figure 7.4). The terminals of these input cables have ionic channels that are mechanically anchored to skin cells through microscopic molecular bridges. Skin movements stretch these physical connections, literally yanking the ionic channel open, once again causing a change in electric potential.

A similar mechanism of transduction is utilized by auditory neurons in the inner ear. These neurons are characterized by hair-like extensions that are mechanically deflected by sound waves, triggering the opening of ionic channels just as on the skin. Olfactory neurons in the nose and gustatory neurons on the tongue are the easiest to understand because their job is to detect the presence of specific molecules, exactly what synaptic receptors must do with neurotransmitters in all standard synapses.

Motor neurons are located in the spinal cord and project their axons all the way to muscles. Therefore, these axons (like the "dendrites" of skin receptors) must be shaped so as to fit into their appropriate part of the body, such as an arm or a leg (figure 7.4). Moreover, when that body part moves (in response to spikes coming down those axons!), so do the nerves inside it. The connection between motor neuron and muscles, called the *neuromuscular junction*, uses classic synaptic communication. This means that muscles have postsynaptic receptors similar to those found in neuronal dendrites. In this case, however, neurotransmitter binding

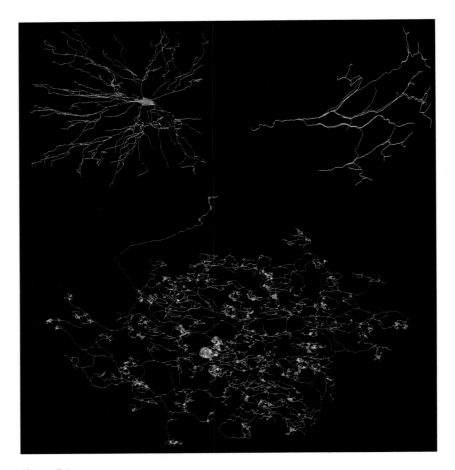

Figure 7.4

Actors and sensors. (Top left) A motor neuron from the mouse spinal cord (data from the author[21]). The soma and dendrites are green (with red bifurcations); this reconstruction contains only the initial portion of the axon (yellow), including a side branch that sprouts off to innervate the same region invaded by the neuron of origin (rendering by Amina Zafar in the author's lab). The rest of the axon (not shown) extends all the way to the muscles of the hind limb. (Top right) The terminal portion of a different motor neuron from the mouse spinal cord.[22] Branches are highlighted with distinct colors to show their peculiarly parallel organization (rendering by Uzma Javed in the author's lab). The dendritic arbor and soma of this neuron were not reconstructed, but their appearance would be akin to that of the neuron shown on the top left. Likewise, the axonal terminal of the neuron on the top left, if it were reconstructed, would appear similar to this tree. (Bottom) Terminal tree of a sensory neuron innervating the skin.[23] Bifurcations are brightened up for better visualization (rendering by Namra Ansari in the author's lab). These reconstructions are freely available online at NeuroMorpho.org (branch thickness was increased to enhance contrast).

triggers coordinated molecular movements altogether resulting in muscle contraction. The dendrites of motor neurons receive their inputs from long axons coming straight from pyramidal cells in the neocortex!

It should not come as a total surprise that the neurons in direct communication with the environment are so special. After all, their function is highly specialized and qualitatively distinct from that of most neurons in the brain, which only send and receive information to and from other neurons. Most if not all brain regions involved in sensory input or motor output also contain other peculiar types of neurons that are not found in other parts of the nervous system. For example, amacrine cells, a diverse collection of neurons that can be distinguished in several dozen subtypes, are exclusively found in the retina even though they are not directly involved in light transduction and only communicate with other neurons. The same is true for the neurons responsible for transmitting signals from the eye to the rest of the brain, ganglion cells, whose axons form the optic nerve. At the same time other neuron types in the retina, such as horizontal cells and bipolar cells, resemble neurons in other parts of the brain both functionally and structurally.

Unique region-specific neuron types are also found in deeply internal parts of the brain that only communicate with other brain regions and not with the rest of the body or the external world. Earlier we mentioned that this is indeed the case for Purkinje cells, the principal neurons of the cerebellum, with their nearly flat but exuberant dendritic arborization. Other types of neurons synapsing onto Purkinje cells are also not typically found elsewhere in the brain. Certain neurons in the medulla, for instance, extend their axons up to the Purkinje cell somata, where they start wrapping around the dendrites like vines, taking the name of "climbing fibers" (refer back to figures 2.4 and 4.3). In another type of neuron unique to the cerebellum called cerebellar granule cells, "T-shaped" axons split into opposite directions at right angles after a short vertical rise and form parallel fibers running perpendicular to the dendritic trees of Purkinje cells.[24] At the same time other neurons in the cerebellum, such as chandelier cells and basket cells, are also common in many other brain regions, including the neocortex.

The number of distinct neuron types in the brain is not yet known, but it is likely that only a minority of the existing types have been identified to date.[25] The rodent brain can be parceled into at least 500 identifiable regions.[26] Although the neuronal compositions of some regions are studied much more in depth than those in others, available data suggest that there may be between one and several dozen distinguishable types of neurons in each region.[27] If neurons in separate regions are always considered different, this amounts to some 10,000 neuron types in the nervous systems of even simple mammals! Rather than providing a long descriptive list of the few hundred neuron types so far characterized in neuroscience labs around the world, we instead attempt to summarize a few distinctions generally applicable across the board to interrelate the observed diversity of neurons with their computational functions. In the next chapter we further sample neuronal diversity by considering a few additional types.

7.4 Protagonists and Supporters

Let's take a moment to review conceptually the functional architecture of the massive human brain network based on the description so far. Each of these hundred billion neurons continuously integrates the signals received from tens of thousands of other neurons, processing this input computationally to generate patterns of spikes. Every neuron transmits its spikes through extensive axonal arbors, communicating the represented messages across tens of thousands of synapses to other neurons. Some neurons transduce information from the external world via sensory receptors primarily in the eyes, ears, skin, tongue, and nose. Other neurons contact muscles to exercise an influence outside of the brain. The vast majority of neurons, however, process information internally, only communicating with other neurons.

As you read this book, neurons in your retina translate pixels into spikes, igniting a cascade of neuronal activity within your brain. Electric patterns in your visual cortex represent the set of pixels in their proper relative positions and sequences, which are recognized as words in the language area. These words are then encoded as more abstract conceptual representations (also in the cortex). In parallel, your mind is probably retrieving thoughts related to the concepts you're reading about as well as instantiating meta-representations such as the feeling of excitement for intellectual understanding. The flow of these activity patterns encoding for the mental state of your present experience also modifies the strength of synaptic connections, even creating some new synapses and eliminating others. As a consequence, you might remember this explanation tomorrow morning even if you would have likely been unable to think of it yesterday. Quite seamlessly, motor commands are also constantly issued to wrap your eyes around each line of text, to flip pages, to take notes, to scratch your head, to blink, and so on.

This view of the brain focuses on the neurons that encode the actual content of the information being processed: visual input, the meaning of words and sentences, the relationship between the represented concepts and previous knowledge, and the coordinated muscular activation required to yield specific body movements. In the cortex the neurons whose activity represents information encoding are indeed the majority—but not the totality. A substantial fraction of neurons (10–20%) have a different role, namely to keep the circuit working.

To understand why these supporting neurons may be necessary, consider the organization of a restaurant as a metaphor. The "inputs" of this structure are raw food and hungry customers. The outputs are satiated customers, financial revenue, and garbage. The processing pipeline can be schematized as the following: waiters welcome and seat customers, relaying their orders to the kitchen, where a chef cooks with the help of an assistant starting from ingredients bought by a purchase manager. Waiters bring the food to the customers and bring their used plates to the dishwasher operators for cleaning. A cashier-accountant collects the payments, keeps the purchase manager funded, and logs the revenues.

The described personnel staff the essence of the restaurant mission but could not by themselves maintain a smooth operation. Somebody else has to sweep the floors at night, pay everyone's salaries and government taxes, turn the lights on and off every morning and evening, change the burnt light bulbs as needed, advertise in the newspaper weekly, and so forth. These activities are not directly relevant to the core of the restaurant process but are nonetheless essential to its proper running. Similar needs are shared with many other businesses such as a barber shop or a consulting firm.

The chef and waiters of the restaurant, the barber in the barber shop, and the consultants in the consulting firm are the staple representatives of each of these organizations; their activity directly fulfills the intended purpose of each company. They are like the protagonists of a story around whom the plot unfolds. In contrast, the cleaning crew and the human resources department ensure that the chef and the barber can perform their activities. They play a supporting role, setting the stage for the central action.

Such a distinction is believed to apply to neurons in the brain just as well. Certain "protagonist" neurons, in some context known as *drivers*, directly represent and process information. Other "supporter" neurons, sometimes called *modulators*, enable drivers to function properly, for example, by maintaining the network activity synchronized to a common rhythm and ensuring that all neurons receive a balanced input of excitation and inhibition.[28] The activity of protagonists, not of supporters, corresponds to mental states. Without supporters, however, some protagonists would fire too much, others too little, and most protagonists would quickly fall victim of incorrect timing of communication both on their dendritic input and axonal output ends. Very soon, the entire network would become unintelligible (and unintelligent). In section 4.1 we compared neural networks to symphonic orchestras, neurons to individual instruments, spikes to their notes, and activity patterns to the symphony itself. In this metaphor, protagonist neurons correspond to the instruments playing the melody (usually strings and winds), whereas supporter neurons would include the very different roles of both percussions and conductor. The distinction, however, should not be construed as a ranking of importance, as all roles are ultimately essential for a properly executed performance.

In the brain as in a symphonic orchestra, the neurons or instruments "playing the content" constitute a majority. Both to reflect this numerical abundance and their main roles in information processing and transmission, the protagonist neurons are often referred to as *principal cells*. The principal cells of the cortex are *pyramidal neurons*, so called originally because of the triangular appearance of their soma.

In addition (and related) to their soma shape, pyramidal neurons are most commonly recognized from other nerve cells because of the form of their dendrites. The trademark dendritic arbor of these neurons is composed of an apical tree stemming out of the tip (or apex) of the somatic pyramid and several basal trees coming out of the opposite side (the base). Despite their characteristic somatic and dendritic profiles, the most distinguishing structural feature of pyramidal neurons is in fact their axonal arborization. The axons of pyramidal neurons can span the entire surface of the brain, and they constitute the backbone of cortical connectivity. If two cortical regions are connected, such as the auditory cortex and the area devoted to language comprehension or the sensory and motor cortices representing the right thumb, it's because their respective pyramidal cells are connected.

Pyramidal neurons are located in various layers or depths of the neocortex but are particularly abundant in layers 2 and 3 (toward the surface of the thin cortical mantle) and in layer 5 (near the bottom). Layer 2/3 and layer 5 pyramidal neurons have different connectivity patterns, but even within a given layer, the axons of different pyramidal neurons can project to different areas. Indeed, the vast majority of cortical pyramidal neurons project most of their axons far from the space invaded by their dendrites. The dendrites typically extend just 2 millimeters, but the axons may reach a foot away from the soma. For this reason pyramidal cells are also often called *projection neurons*, although in many cases they *also* extend copious ramifications within the local volume containing their soma and dendrites.

Other neuron types mostly or only establish outgoing connections with neurons in their vicinities and are typically called *interneurons*. The axons of interneurons do not reach distances significantly farther that those spanned by their dendrites. Nevertheless, in cortical interneurons the axonal tree is typically denser than the dendritic arborization, so the total branching length of the axon may still be several orders of magnitude greater than that of the dendrite. Moreover, even though the volumes enclosing the interneuron dendrites and axons are similarly sized, the axonal and dendritic spaces of the same interneuron may totally coincide, only partially overlap, or be completely disjoint, with the axon on one side of the soma and the dendrites on the other. Importantly, interneuron connectivity is only local in terms of their output, not their input: their dendrites, like those of principal cells, receive axonal contacts from both local and projection neurons, the latter being located very far away (figure 7.5).

In the cortex the distinction between principal cells and interneurons is sharpened by an important observation, which may prompt some readers to temporarily flip back a few dozen pages for a quick review of section 3.2. The axons of pyramidal cells release the excitatory neurotransmitter glutamate, whereas those of interneurons release the inhibitory neurotransmitter GABA. As a consequence, in the postsynaptic sites contacted by pyramidal cells, incoming signals open sodium channels, causing an influx

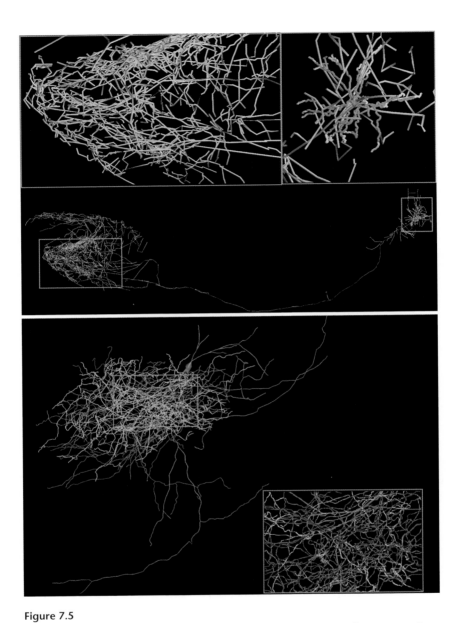

Figure 7.5

Projection neurons and local interneurons. (Top) Layer 2 spiny stellate neuron from the rat entorhinal cortex (see note 4 in chapter 4). The left and right insets zoom in, respectively, on the distal axonal tree, colored blue, and on the star-shaped dendrites,[29] colored green (with yellow terminals), emphasizing the long axonal projection in between (rendering by Uzma Javed in the author's lab). (Bottom) CA3 interneuron from the rat hippocampus.[30] The soma and dendrites are colored blue; the dense but local axonal arbor is colored in shades of green, with the inset zooming in on the dense branching (rendering by Namra Ansari in the author's lab). Neurons of this type receive connections from the type of neurons depicted in the top panel. Both reconstructions are freely available online at NeuroMorpho.org (branch thickness was increased to enhance contrast).

of positive charges, whereas in those contacted by interneurons they open chloride channels, causing an influx of negative charges.

Firing in pyramidal cells thus increases the probability of triggering a spike in the neurons they contact. In contrast, when interneurons fire, they tend to silence the neurons they contact. This distinction, like the local or nonlocal character of the connectivity, also applies only to the output, not the input, of pyramidal cells and interneurons. On their dendrites, all cortical neurons receive a mix of excitatory and inhibitory signals and contain receptors for both glutamate and GABA.

The fact that pyramidal cells communicate with other brain regions to stimulate activity, whereas interneurons only talk to neighbors to keep them quiet, suggests a division of labor between protagonists and supporters. In this view excitatory neurons would be the protagonists of cortical computation, representing, processing, and transmitting the content of information. Inhibitory neurons would provide a support role, maintaining the proper tempo of activity and fine-tuning network dynamics. This functional distinction between pyramidal cells and interneurons, although constituting an extreme oversimplification, is nonetheless convenient to foster speculation on their computational roles.

Much of our reasoning on the brain-mind relationship (chapters 4–6) appears more directly relevant to principal neurons than to interneurons. However, it would be a mistake to discount the role of interneurons from encoding mental content. The first principle of the brain-mind relationship posits a correspondence between mental states and spatial-temporal activity patterns, which are the combined outcome of both excitatory and inhibitory action. Interneurons can completely alter both the spatial and temporal patterning of the network. Therefore, they might ultimately affect the mind as much as the principal cells. Moreover, the same physical constraints linking axonal-dendritic proximities to the capability of forming synapses apply broadly to all neurons and to excitatory and inhibitory contacts alike. We provide a few illustrative examples of specific supportive functions of different interneuron types in cortical circuits in the next chapter.

A few additional characteristics distinguish pyramidal cells from interneurons, as summarized in table 7.1. The peculiarly shaped dendrites of pyramidal cells are systematically aligned along the depth of the cortical sheet, always with the basal trees toward the deeper layers and the apical

Table 7.1
Properties of principal neurons and interneurons in the cerebral cortex

Property	Principal neurons	Interneurons
Abundance	Numerous	Sparse
Dendritic shape	One or few main types	Highly diverse
Dendritic surface	Spiny	Typically aspiny
Axons projection	Long distance	Local
Neurotransmission	Glutamatergic	GABAergic
Target on principal neurons	Dendritic spines	Dendritic shaft, soma, and axon
Postsynaptic effect	Depolarizing	Hyperpolarizing
Postsynaptic function	Excitatory	Inhibitory
Network function	Drivers	Modulators
Network role	Protagonists	Supporters

tree extending to the outer surface. The dendritic arbors of cortical interneurons, in contrast, come in many different shapes and any orientation. The dendrites of pyramidal cells are covered with tens of thousands of tiny protuberances, dendritic spines, where all of the excitatory synapses and very few of the inhibitory ones are formed. Although a minority of interneurons can be spiny, most of them have smooth dendrites receiving both excitatory and inhibitory input. The axons of pyramidal cells contact their postsynaptic targets on the spines if present or on the dendrites of aspiny interneurons. The axons of interneurons, in contrast, can contact several different postsynaptic locations (as exemplified in the next chapter).

This dichotomy between principal cells and interneurons is far from airtight, and several neuron types fall through the cracks. Some GABA-releasing inhibitory neurons have far-projecting axons, and some local interneurons are excitatory and release glutamate. It is not yet known whether either of these two intermediate classes plays a role of protagonist, supporter, somewhat intermediate, or something different altogether. Similarly, there are many exceptions to virtually all of the entries of table 7.1, although each of the listed properties tends to apply in the majority of cases.

In a broader sense the distinction between principal cells and interneurons should not be overgeneralized. Much of our data, knowledge, and examples are from the cortical regions (neocortex and hippocampus) of

Figure 7.6
Mood, trees, and neuromodulation (both panels modified by the authors from photographs by Daniel Segrè).

adult rats, mice, and, to a lesser extent, other mammals (more noticeably monkeys). Although many of these principles also apply to other organisms, developmental stages, and brain regions, the specifics of network organization vary considerably in all these cases. Certainly in invertebrates (which constitute, remember, the majority of existing species with a nervous system!) there is no identification of drivers or protagonists with projecting excitatory pyramidal cells and of supporters or modulators with local inhibitory interneurons. Even in mammals the principal neurons of noncortical brain regions are not pyramidal cells and do not necessarily release glutamate. For example, Purkinje cells in the cerebellum release GABA and are inhibitory! Moreover, the rules are different early in development even in the cortex, when GABA-releasing interneurons may exercise an excitatory effect, with mechanisms and consequences that are still much under debate in neuroscience.

Even in the mature cerebral cortex of rodents and humans, there are important neurons that release neurotransmitters other than glutamate and GABA, such as serotonin, dopamine, norepinephrine, and acetylcholine, resulting in neither systematic excitation nor inhibition. Instead, these different neurotransmitters exert modulatory functions such as amplifying or dampening other synaptic signals (whether excitatory or inhibitory) and controlling plasticity via receptors that activate powerful chemical cascades inside the postsynaptic neurons instead of opening ionic channels.[31]

From the cognitive point of view, these modulatory neurons do not directly code for information content but rather define the state of mood, attention, arousal, and motivation. Although functionally these neurons are clear supporters, their soma is outside of the cortex, and their axons travel a long distance to reach their target, spreading widely to project throughout broad regions at once. Malfunctioning of these neuromodulatory systems is implicated in pathological conditions as diverse as depression, obesity, schizophrenia, and addiction. The neuronal machinery involving these neurotransmitters is targeted by a broad variety of drugs, from caffeine to alcohol, from cannabis to cocaine, from Ritalin to Prozac (figure 7.6).

8 Neuron Types

8.1 All Neurons Are Different, but Some More Than Others

The last chapter and previous parts of the book introduced many aspects of neuronal diversity in the brain. At one extreme every neuron has a unique axonal and dendritic shape that is not repeated in any other neuron, in the same region and organism or otherwise. At perhaps the opposite end of abstraction, we distinguished information-carrying protagonist neurons from their supporting staff of modulatory interneurons. Protagonists and supporters tend to have different arbor geometries and other dissimilarities (see table 7.1). There are, however, many intermediate levels of diversity worth considering.

Imagine taking a leisurely stroll throughout the brain forestry just to admire the beauty of neurons without regard for their computational function in the network. While certainly appreciating the individuality of each tree, you would also be able to recognize clear similarities among groups of neurons. Some trees would look like birches, others like willows, still others like pines. Many types of trees would be apparent, with many neurons identifiably in each type (figure 8.1).

We have already encountered several such neuronal "tree types," such as pyramidal cells, Purkinje cells, and granule cells. Continuing our promenade in the garden of the brain, we walk by a conglomerate of densely intertwined branches looking like a nest. Peering inside the nest, rather than bird eggs, we find the soma of another neuron. The branches forming these nests are from the axons of neuron types called basket cells, precisely because they form baskets around the cell body of the neurons they target. Just as we finish admiring the basket cell, we notice nearby another tree of a completely different shape: that of a candelabrum so spectacularly rich

Figure 8.1
A botanic garden of shapes (Daniel Segrè, 2013, Mount Auburn Cemetery, Cambridge, Massachusetts).

that, if its arms were real candlesticks, it would hold thousands of burning flames. This is the axon of a chandelier cell, again named after the peculiar shape of its tree. Each of its candle-like terminals targets a different postsynaptic neuron, making a handful of adjacent synapses along the stick.

Basket and chandelier cells are found in different regions of the brain, including the neocortex, the hippocampus, and the cerebellum. Assuming we entered the neuronal park from the cortical gate, let's explore a few more trees that have been characterized there.[1] Among an abundance of pyramidal cells, one beautiful arbor displays an impressively tight mesh of very thin fibers, the axons of neurogliaform cells (figure 8.2). A similar closed-knit tree, but spread thorough a broader volume, is also seen in the hippocampus, where it gives its name to the ivy cells from which it originates due to the familiar look of the homonymous plant. Another peculiar shape in the cortical wood is that of double bouquet neurons, named after their axonal arbor, which looks like two bunches of flowers held in one hand, one kept right-side up and the other upside down. Another neuron, named after its discoverer Martinotti, has a very long axonal stem rising vertically to the top cortical surface where it finally sprouts into an exuberant horizontal mesh.

Each of the neuron types described above, from basket cells to Martinotti cells, is distinguished for the peculiar shape of their axon. Indeed, because of its sheer length, the axonal arbor often offers immediate recognition cues. Several neuron types, however, are most easily characterized by the shape of their dendrites, including the famous cases of pyramidal cells, with their basal and apical trees, and of Purkinje cells with their highly complex, dense, and flat branching. Whereas Purkinje cells are exclusively found in the cerebellum, different types of pyramidal cells are located all throughout the cortex or outside, each with distinct axonal trees. For example, pyramidal cells in the motor cortex project their axons to the spinal cord, and pyramidal cells in the auditory cortex project their axons to language-processing areas (among other places). Some of the neurons named after their distinctive axonal trees, such as basket cells, do not have uniquely shaped dendrites—instead, their dendritic arbors can come in several different shapes. For example, chandelier cells can have horizontally oriented dendrites or pyramidal-looking apical and basal trees.

Other examples of neurons named after their dendritic structures are those of the hippocampal mossy cells and of stellate cells. Mossy cells

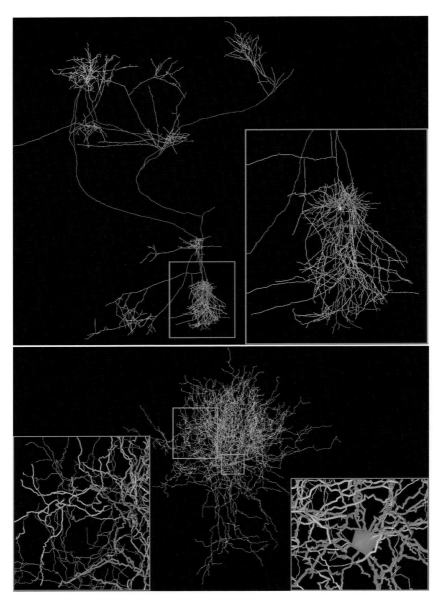

Figure 8.2
Have axons, will travel. (Top) A pyramidal neuron from the cat visual cortex.[2] The dendritic tree is colored purple (magnified in inset); each subtree of the axonal arbor is colored differently depending on the invaded regions (e.g., recurrent collaterals also visible in inset are green). (Bottom) A neurogliaform interneuron from the rat motor cortex (see note 21 in chapter 3). Soma and dendrites are colored pink, axon light blue; the left and right insets zoom in on details farther from and near the soma, respectively (renderings by Uzma Javed in the author's lab). Both reconstructions are freely available online at NeuroMorpho.org; branch thickness was increased to enhance contrast.

are identified by the thorny excrescences covering their dendrites, which serve a bit like giant conglomerates of spines. Stellate cells, so called because of the star-like arrangement of the dendritic trees, are found both in the cerebellum and in the cortex. In the latter they come in two varieties: one that is spiny and excitatory and another that is aspiny and inhibitory. These distinct types of stellate cells can also be distinguished from each other (as well as from their cerebellar cousins) by their different axonal shapes. A particular kind of interneuron only found in the retina is called a "starburst" cell also because of the shape of its dendrites.

Other neurons have equally picturesque and uniquely identifying dendritic and (especially) axonal structures but for historical reasons have been named after the shape of their cell body. These names are usually a legacy of the early days in neuroscience before the discovery of chemical preparations to visualize dendrites and axons, when somata were the only visible parts of the neurons. We mentioned earlier various types of granule cells, named after their small, tightly packed somata, found in the hippocampus, cortex, cerebellum, and olfactory system, and of course the name of pyramidal neurons comes from the form of their cell body. Other peculiar cases are those of mitral cells and spindle cells, whose somata are shaped respectively like a bishop's hat and like a tapered spinning rod.

A classic outdoor activity is to look up at the clouds and call out shapes. As a child, I enjoyed the night version of this game, connecting the stars with imaginary lines when the sky was clear to draw new constellations in my mind. In contrast with the fleeting nature of ever-changing clouds, the steady reliability of the stellar lineups allowed me to identify the same shapes from one night to the next (for a scientist-to-be, reproducibility is a must). Never finding the Big Dipper sufficiently fulfilling, I invented in turn the Giant Strawberry, the Cosmic Donut, and the Intergalactic Ice Cream Cone. Now in the neuroscience lab decades later, I still often indulge in the same playful habit, calling out amusing shapes when I inspect neurons under a microscope. Most recently, as I was finalizing the illustrations for this book, my oldest son masterfully outperformed me when he claimed that the tree in the picture on my desk looked like a chicken (we did have roasted chicken for dinner that night). Younger brothers gathered around, rapidly triggering the discussion of whether neurons in the chicken brain look like chickens. Although I did not need to look up the data to provide a negative answer, I nonetheless felt compelled to check out what chicken neurons look like (figure 8.3).

Family interludes aside, let's return to the structural identification of neuron types. Additional anatomical considerations differentiating neurons include their size, orientation, and the already mentioned presence of spines on the dendrites of some neurons. The size of the soma can vary considerably, often in parallel with the extent of the dendritic tree. The most egregious cases include giant multipolar neurons in the spinal cord, giant pyramidal cells (also called Betz's neurons) in the neocortex, and giant radiatum neurons in the hippocampus. The soma of these neurons can have a diameter ten times larger than that of granule cells.

Still the extent of axons is often even more variable from neuron type to neuron type. It has been estimated that a sensory axon responsible to detect touch at the tip of the tail in the largest dinosaur may have reached across more than 170 feet,[4] making it the longest individual cell in the history of the biosphere.

The orientation of axonal and dendritic trees is perhaps even more telling. Classic pyramidal cells, for example, are oriented vertically in the cortex, with the apical tree rising to the top surface and basal dendrites pointing the opposite way. Pyramidal-looking neurons oriented the other way around (apical down, basal up) or in a perpendicular (horizontal) direction constitute two different (and rarer) neuron types. More generally, neurons with a dendritic and/or axonal polarity (i.e., displaying a preferential direction of growth) can be described based on the alignment of their main axis relative to the anatomy of the brain. Similarly, when trees are flat, they can be classified on the basis of the rotation of the two-dimensional plane they lie in relative to the surrounding tissue.

All of these visual descriptors responsible for "the looks" of every neuron can be quantified by rigorous geometric measures. For example, the total length of the axonal and dendritic cables, the average length and diameter of each branch, the orientation angle of the entire tree, the density of spines, and so forth can each be measured precisely in each neuron, producing a long series of numeric values.

Imagine, for instance, recording for every neuron the ratio between axonal and dendritic spans. In practice this ratio will never be exactly the same in any two neurons. However, neurons of the same type will tend to have more similar values than neurons of different types. In general, projection

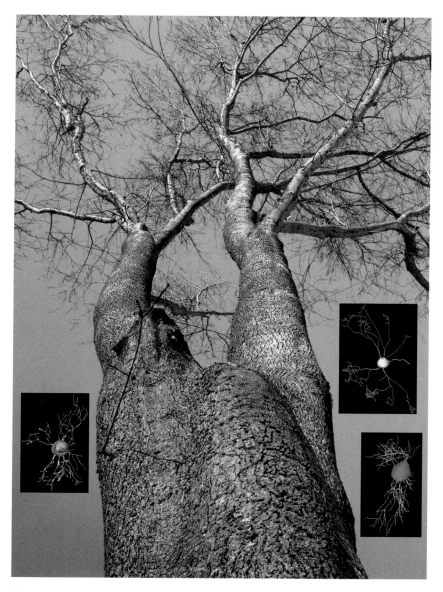

Figure 8.3
Chicken tree and chicken neurons. Three bipolar neurons from the chicken brainstem[3] overlaid on top of "The geometry of the possible" (Daniel Segrè, 2014, Fresh Pond Park, Cambridge, Massachusetts). Neuron renderings (clockwise from left) respectively by Uzma Javed, Amina Zafar, and Namra Ansari, with help from Todd Gillette, from the author's lab. Bifurcations and/or terminations are highlighted in different colors. The reconstructions are freely available online at NeuroMorpho.org; branch thickness was increased to enhance contrast.

neurons will have much greater numerical values for this measure than interneurons. Now consider measuring another parameter, such as the average length of dendritic branches. Again, the value logged by a Purkinje cell will be typically closer to that of another Purkinje cell than to that of a spinal motor neuron, which instead will have a similar branch length as other spinal motor neurons.

If you were to build a diagram plotting these two dimensions against each other (say, ratio between axonal and dendritic spans on the horizontal X axis and branch length on the vertical Y axis), you could record a point for every neuron you measure. After dotting hundreds of points in this graph, neurons of the same type would tend to group (or to *cluster*) together. This same process can be repeated with a third type of measure, yielding three-dimensional "clouds" of neurons, and in more abstract ways with any number of dimensions. Although no two neurons ever fall in the exact same spot in these quantitative diagrams, neither do their scatters cover the entire geometric space uniformly. Instead, they form a limited number of groups. This is indeed the signature of neuron types: each cluster represents a neuron type, and points within a cluster belong to the same type. The same happens with botanical trees.

There are also more qualitative features of axonal structure that differ among neuron types. The axons of all pyramidal cells and of some interneuron types contact their dendritic targets on the spines if they are present. Some interneurons connecting to spiny dendrites, in contrast, prefer to establish their synapses directly on the shaft of dendritic branches (in between spines). Some neurons, such as basket cells, make multiple contacts on or around the cell body of their postsynaptic neurons. Others, including chandelier cells, target the initial segment of the axonal arbors of the neurons they contact, also with several synapses in row. Interestingly, only inhibitory synapses are found on the soma or axon. These axosomatic and axoaxonic synapses are different from the typical axodendritic contacts that make up the vast majority of brain connectivity.

Differences in synaptic locations are important because of the considerable computational consequence. Every synapse on the dendrite is "counted" more or less in the same manner: some may be stronger, others weaker, some are farther in the tree, some closer, but the information each of them transmits is integrated and conveyed to the soma. In contrast, synapses at the soma or on the axonal initial segment can potentially alter the result of the entire dendritic computation. The dendritic tree continuously combines input from many hundreds of excitatory and inhibitory synapses that happen to be active at any given time (out of the tens of thousands that are physically present on the tree). If the dendritic arbor relays a "go" sign to the soma, meaning that excitation is dominant and the neuron should spike, synapses from basket cells on the soma can still silence the neuron. And even if a spike is triggered, synapses on the axonal initial segment from chandelier cells can issue a "just-in-time" veto blocking the spike propagation down the tree. Because its axons contact many hundreds of pyramidal cells, when a single chandelier cell fires, it stops spiking in a substantial portion of the network.

8.2 There's More Than Meets the Eye

At this point in our story the characteristic shape of the axonal and dendritic arbors of different neuron types should not be all that surprising because neurons use axons and dendrites to form interconnected networks. If various types of neurons need to fulfill their own distinct connectivity requirements with other neurons, they need appropriate inputs and output trees. The ultimate mission of neurons, however, is to communicate with other neurons throughout the circuit. This communication is mediated by sequences of axonal spikes. Although all neurons share the same language of spikes, the particular "voice" and "dialect" vary from one neuron type to another.

Some neurons spike faster (and are called, well, fast-spiking), and others more slowly (sometimes referred to tonic-spiking), much as some people speak faster and others slowly, but there are also neurons that just spike at irregular intervals. Some neurons emit sequences of very fast spikes followed by periods of silence before the next sequence (bursting cells), a bit like those people who speak really fast but then take long pauses between sentences. When these bursts and intervals are irregular, the neurons are called stuttering cells. Some neurons spike fast when they start firing but then progressively slow down spike after spike (called adapting cells), just as when someone begins uttering an opinion at full speed, only to realize right after starting the sentence that the idea requires more thought.

Although all spikes are fundamentally similar, in some neurons the amount of electric signal is progressively reduced when multiple spikes are fired in rapid sequence (attenuating cells).

More generally, when a neuron receives a signal that continues in time, such as a lengthy barrage of synaptic excitation, its firing response can be understood in terms of an initial phase, called the transient, and a stable condition, called the steady state. For instance some neurons have bursting transients followed by tonic spiking as steady state. In the so-called delayed or late-onset neurons, the transient simply consists of a period of silence (absence of firing) prior to the steady state. Other neurons lack a transient response altogether and immediately enter their steady state. The threshold to fire also depends on the neuron type; some neurons require less excitation than others to trigger the first spike.

Recall that the duration of a spike (and that of the minimum interspike interval) is approximately 1 millisecond, whereas mental states last no less than 50 milliseconds. Thus, a message from one neuron to its (tens of thousands of postsynaptic) recipients can easily include several dozens of spikes.

On the input side the dendritic response to synaptic stimulation also varies among neuron types. To be more precise, the synaptic response can depend on the neuronal identity of the presynaptic axon, of the postsynaptic dendrite, or both. For instance, some synapses transmit longer-lasting signals than others. The amount of electric signal depends on the neuron types as well but can also be adjusted based on activity and experience (synaptic plasticity).

Moreover, when a synapse receives two or more spikes in rapid sequence, such as from a fast-spiking, bursting, or stuttering cell, the second and subsequent signals it transmits may not necessarily be equivalent to the first: they can be progressively stronger (facilitating synapses) or weaker (depressing synapses). Furthermore, various neuron types have different propensities to fire at specific phases of the overall network activity (these are the rhythms detected by electroencephalography): some neurons prefer to spike at the peak of activity, others at the trough; among the neurons that spike at intermediate activity levels, some have a preference for when the overall activity is decreasing, others for when it's increasing.

The distinguishing input and output electric signature of each neuron type is due to the specific distribution of ionic channels and neurotransmitter receptors on its axonal and dendritic arbors. These biochemical components are made out of particular proteins, each coded by a certain gene. Every single neuron in the brain, and in fact every cell in the body, contains the entire genome of the organism. These genes include all the information necessary to code for every protein any cell of that organism will ever produce. Every cell, however, and every neuron in particular, only expresses a subset of the ~25,000 genes from the genome, yielding a unique portfolio of proteins.

In neurons proteins are the molecular constituents of ionic channels and neurotransmitter receptors on the axonal and dendritic trees. The subset of expressed proteins therefore determines the distribution and composition of the biophysical machinery giving rise to the each neuron's specific *computational signature*. As a result, the molecular fingerprinting of every neuron guarantees an exquisitely determined propensity to fire spikes in particular patterns. No two neurons express the exact same proteins in identical proportions. However, much as for axonal and dendritic shapes, clear groupings still emerge.

Neuron-specific gene expression patterns underlie all functions of the neuron, not just spiking and postsynaptic reception. For example, different neuron types can release different neurotransmitters. Specific molecular machineries must be in place in order to produce glutamate, GABA, dopamine, acetylcholine, serotonin, or other such chemicals. This is a fundamental aspect of the biochemical characterization of neuron types because the identity of the neurotransmitter crucially determines the excitatory, inhibitory, or modulatory effect of each and every spike fired by the neuron on all the tens of thousands of postsynaptic targets! Other neuron type-specific molecular markers (with technical names such as parvalbumin or somatostatin) involve more subtle aspects of cellular function that are not yet completely understood.

Every neuron is characterized by a complex combination of axonal-dendritic shape, spiking pattern, and biomarker expression. For example, basket cells of one type are fast-spiking and contain parvalbumin, whereas Martinotti cells are stuttering and contain somatostatin (figure 8.4). The relationship and interactions among these different aspects of neuronal identity are not yet well understood and remain the focus of intense neuroscientific research.[6] Moreover, each neuron type has its own developmental origin, which is specified by the precise moment it begins to differentiate,

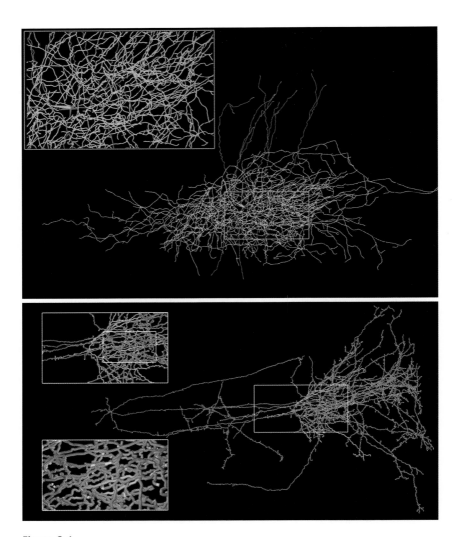

Figure 8.4
Beyond the appearance of neuron types. (Top) A cortical basket cell[5] with dendrites colored magenta, axon yellow, and inset zooming in on arbor details. This type of neuron is characterized by the molecular expression of parvalbumin and a fast-spiking firing pattern. (Bottom) A cortical Martinotti cell (see note 3 in chapter 2) with dendrites colored blue, axon magenta, and progressively zoomed-in details in the insets. This type of neuron is characterized by the molecular expression of somatostatin and a stuttering firing pattern (renderings by Amina Zafar in the author's lab). Both reconstructions are freely available online at NeuroMorpho.org; branch thickness was increased to enhance contrast.

as well as its migration trajectory (see section 7.2). Furthermore, neuron types differ in their plasticity properties: some neurons are more prone than others to modify their characteristics in response to experience and activity. Plastic changes that may vary among neuron types include synaptic strength in outgoing (axonal presynaptic terminal) and incoming (dendritic postsynaptic receptors) connections, intrinsic electrophysiological properties (e.g., excitability), and even structural malleability, such as the rate at which spines can twitch in and out of dendrites, varicosities can crawl up and down the axons, and even entire branches can grow, retract, or bifurcate.

8.3   Location, Location, Location!

We have seen as early as in section 1.2 that the brain is composed of many distinct regions, each specialized in particular cognitive and computational functions, such as the hippocampus for autobiographic memory processing and spatial navigation, the cerebellum for fine coordination of movements (and perhaps for coordinating thoughts, too), the amygdala for emotional learning, and so on. More recently in section 7.1, we have discussed how these separate regions often contain different neuron types.

An echo of this diversity at the whole-brain scale can be observed at a finer level of granularity. Few if any brain regions are internally uniform in terms of either structure or function. On the contrary, each brain region is typically organized into subregions. The best known case is perhaps that of the neocortex, which has clearly separate areas encoding for each sensory modality (vision, audition, etc.), action planning, motor execution, language comprehension, and more. Although not as well understood, the cerebellum and thalamus appear to be similarly organized. The division of brain regions into subregions does not always follow the modality of represented content (sight, sound, touch, etc.). Some regions process information with a series of computational transformations each carried out in a separate subregion. This is, for example, what happens in the hippocampus.

The hippocampus operates to store certain experiences into long-term memory and to retrieve them as needed. By experiences we mean *episodes* lived in the first person with full awareness. For example, your hippocampus would activate if you were asked to remember the last time you ate with chopsticks but not if you were asked to explain what chopsticks are or to give a demonstration of their use. The hippocampus is also involved with representing so-called *prospective memories* or intentions about the future, such as when you remember to pick up Chinese take-out (and chopsticks!) on the way home from work.

The hippocampus furthermore encodes a spatial map of the environment[7] and is particularly active when planning a route or following a path to a destination. Somehow, the hippocampus takes the present mental state (the understanding of what's happening, the formulation of a firm intention, or the awareness of one's location) and commits it to a stable representation. Later (sometimes fifty or more years later!) the hippocampus can reinstate that memory, intention, or spatial information as mnemonic imagery.

How the hippocampus works is far from completely understood, but a few pieces of the puzzle have been put together. The hippocampus receives its input from a neocortical subregion called the entorhinal cortex and sends its output back to the same subregion. Therefore, by and large, the entorhinal cortex constitutes the entire universe within which the hippocampus operates. The hippocampus is traditionally known as the *trisynaptic circuit* because the flow of information follows a three-step neural architecture. To start with, the main input from the entorhinal cortex reaches the first hippocampal subregion, named dentate gyrus; next the dentate gyrus sends its signals to the second subregion, called CA3; then CA3 communicates with the third subregion, CA1; and last CA1 projects the hippocampal output to the entorhinal cortex. This description is only a superficial approximation of the actual organization of the hippocampus, but it is sufficient to understand why circuit connectivity fundamentally links the computational roles of neuron types to location within a brain region.

The principal neurons of the dentate gyrus, granule cells, are believed to disentangle highly complex mental states from the entorhinal cortex (whole experienced episodes) into independent information units. Scientists do not yet know much about these information units, but for the purpose of this explanation imagine decomposing the memory of your last Chinese food outing into the General Tsao's chicken you ordered, the chopsticks you ate it with, the downtown location of the restaurant, the pouring rain that night, the friend you were with, and the conversation about career advancement you had with her. The principal neurons of CA3, pyramidal cells, form a densely interconnected network, which might operate to integrate the above information units by linking them together with synaptic connections. As a result the next time your friend asks "remember when we talked about my promotion?" you will recall chopsticks and pouring rain, even if she did not mention them. The principal neurons of CA1, also pyramidal cells, are believed to enable the comparison of the hippocampal and entorhinal representations to determine which aspects of the current experience are novel and which are familiar.

Even if the principal neurons of both CA3 and CA1 are relatively similar in terms of dendritic shape, firing activity, developmental origin, and molecular content, they are still distinguishable on all of these counts (and dramatically different with respect to their axonal trees). Most importantly, they have different connectivity patterns with respect to both their input and output and carry out clearly distinct functions. Thus, they are different neuron types.

The question of whether or not similar neurons in different regions should be considered of the same type is less straightforward for local interneurons. Some interneurons, for example, receive excitatory input from pyramidal cells and provide inhibitory feedback on the same neuron types (not necessarily the same individual neurons). These local interneurons exist both in CA3 and CA1, and they likely exercise the same local function, in this case feedback inhibition. However, in one case they do so in a circuit devoted to information integration (CA3), and in the other they contribute to information comparison (CA1). Some neuroscientists consider them to be the same neuron type, but others prefer to treat them as two different types,[8] fueling a never-ending debate between "lumpers" and "splitters."

I have long indulged in my own fascination with neuronal trees, and those in the hippocampus are my absolute favorites. One Sunday morning several years ago I woke up early and, after picking a shirt from my closet, I started fidgeting with the metal hangers. After untwisting a bunch, I began to coil them back together in the shape of a dentate gyrus granule cell. A couple of hours later my wife got up, peered inside the closet (where I had almost finished bending a 2-foot-tall dendritic tree into shape), shook her head with a sigh, and sent me downstairs to give the kids breakfast. But when I opened the pack of sliced bread I realized that twist ties were even easier to work with than metal hangers. After hastily preparing French toast for four boisterous boys, I proceeded to create two smaller versions of

Figure 8.5
The neuronal circuit of the hippocampal formation: virtual and physical models. (Top) A virtual reality rendering of thirteen individually color-coded neurons from the rat hippocampus (model and image by Todd Gillette from the author's lab). The model includes one principal neuron, complete with dendrites and extensive axonal projection, from each of four subregions (dentate gyrus, CA3, CA1, and entorhinal cortex). A second excitatory neuron is present in each subregion with full dendritic arbors but only the local axonal branches. Last, there are five local inhibitory interneurons with axonal and dendritic trees. (Middle) *Mental Floss* (2012), a copper wire and glass bead sculpture (on acrylic base and Plexiglas supports) by the author and Alice Quatrochi, representing a scaled physical embodiment of the above circuit (photograph by Evan Cantwell, George Mason University). The relative position, orientation, and circuit connectivity of the neurons are all anatomically accurate. (Bottom) A detail on area CA3 (photograph by Evan Cantwell, George Mason University). The glass beads (representing idealized dendritic spines and axonal varicosities) have distinguishing colors for each individual neuron, allowing viewers to visually follow their branches even where densely intertwined. The sculpture is approximately 8 feet wide. The width of a real rat hippocampus is only 2–3 millimeters (roughly 1:1000) and contains more than 1.5 million neurons (over 100,000 times those included in *Mental Floss*). A more complete description of *Mental Floss* with a link to a three-dimensional animation of the virtual model is available on the website of the author's lab at http://krasnow1.gmu.edu/cn3/MentalFloss.

granule cell dendritic arbors with twist ties and finger-warped paper clips. Soon thereafter, I initiated a unique art-science collaboration to create a scaled but full-fledged neuronal sculpture of the hippocampal circuit that is still on display at George Mason University (figure 8.5). The final choice of material was electric wire, which is highly appropriate given the function of neurons.

So far we have considered how the computational function of neuron types depends on the location of their soma—say in the hippocampus as opposed to cerebellum or in CA3 as opposed to CA1. The essence of neurons, however, is in their arbor-like qualities. As we have seen in previous chapters, axons of principal neurons can project to faraway brain regions and multiple subregions, whereas the axons of interneurons and the dendrites of all neurons stay within the local subregion. Nevertheless, even subregions are divided into even finer parcels. In the neocortex and hippocampus, for example, subregions tend to follow a laminar division. In other words they are organized in distinct layers (like certain cakes). Axons and dendrites of different neuron types (both principal cells and

Neuron Types

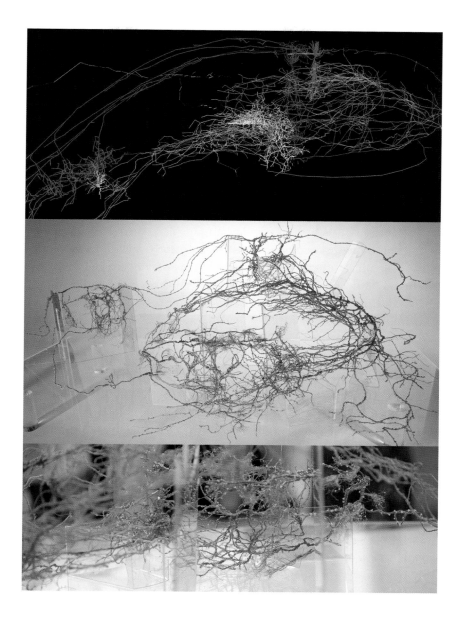

interneurons) can be extremely selective as to the layers they invade. This is very important because the location of axons and dendrites is most telling of circuit connectivity and thus of function. Consider for the sake of example the dendrites of a given neuron type as confined within certain layers—let's call them "superficial" layers relative to the depth of the brain tissue. Imagine that the axons of another neuron type only invade the deep layers. It stands to reason that the former neurons cannot receive any direct information from the latter.

The laminar specificity of axonal and dendritic arbors could be considered a shortcoming of neural architectures. After all, if the axons and dendrites of all neuron types invaded all layers of their subregions, a more varied connectivity repertoire would be available, which arguably might lead to more powerful computation. Unfortunately, such a hypothetical scenario would also increase the amount of necessary neuronal cable, thus resulting in either fewer neurons or larger heads (if you think the latter is a good idea, consider the pain of labor and think again!). Moreover, the compartmentalization of axons and dendrites into layers actually organizes networks into local circuits enabling important computational designs. One such example is the case of feedback inhibition mentioned above: CA1 pyramidal cells have axons traveling in the deepest layer of this region of the hippocampus and dendrites ascending to the most superficial layers, where they collect input from entorhinal cortex. CA1 feedback interneurons have dendrites confined to the deep layer, where they receive stimulation from the axons of pyramidal cells while sending their own axons up to the superficial layer, where they contact the dendrites of other pyramidal cells. Notice how this layering prevents the entorhinal cortex from exciting these interneurons directly, thus preserving their pure feedback role.

Last, we should not discount the position of the soma altogether. Even though axonal and dendritic trees render the connectivity of the network fairly independent of where neuronal cell bodies are located, the location of synapses relative to the soma has important functional consequences for both the input and output of neurons.

Consider, for example, two neurons, both having dendrites in the deep layer and axons in the superficial layer, but with somas in different locations. The soma of the first neuron is positioned in the deep layer, resulting in short dendrites and a long axon. The soma of the second neuron, in contrast, is in the superficial layer, such that the path to the axonal terminals is shorter than that to the dendritic tips. Although these two neurons could receive and send connections from/to the same partners, the incoming signals will arrive near the soma for the first neuron, and farther for the second. Conversely, the outgoing spikes will travel a longer distance to the synapse in the first case than the second.

These geometric considerations affect neuronal computation because dendritic signals can be attenuated, amplified, or integrated with other inputs along the path to the soma. Although the axonal output is all-or-none, longer axons imply greater temporal delays and even risk of transmission failure along the way.

In summary, neurons can be differentiated not only on the basis of their arbor shapes and their physiological or molecular characteristics but also on their physical position in the brain. The old adage regarding the three most important factors when choosing a home seems well suited for neuronal identification: location, location, location! Specifically, the region and subregion in which different neurons are found is intimately tied to their function; their dendritic and axonal laminar patterns respectively determine their input and output connectivity; and the location of the synapses relative to the soma affects signal integration and communication between specific neuron types.

8.4 A Multitude of Multitudes

In our virtual stroll into the neural woodland we have encountered dozens of different neuron types. Numerous others have been characterized since the early days of neuroscience in laboratories all around the world, new neurons continue to be discovered every year, and there is no doubt that many more are waiting to be found. The awe-inspiring diversity of the neuron types that are already known begs the question: how many distinct types of neurons are there in a human brain? This question remains as of this day wide open, but it is nonetheless fundamentally important if we are to understand the brain-mind relationship in physical and biological terms.

As you may recall from the last section, a prominent determinant of neuronal identity is the location of its axons and dendrites in particular brain regions, subregions, and layers. Such crude characterization forgoes many other aspects of neuronal morphology (size, shape, orientation) and all other distinguishing molecular, electrophysiological, and developmental elements we discussed. Nevertheless, the position of its axons and dendrites in a distinct brain location ultimately defines the potential of a given neuron to make input and output contact with the rest of the network. This connectivity with the rest of the circuit is arguably the quintessence of "neuronness." Based on this very simple perspective, we can make a simple back-of-the-envelope computation to estimate the magnitude of the problem in gauging the number of neuron types.

Let's divide the brain into the right and left hemisphere, front and back halves, and top and bottom parts. Admittedly, this two-by-two-by-two split is a laughable caricature of the complexity we described in section 1.2. Yet it's eye-opening to just count how many different neuron types we can define based on such minimalistic parceling. For any neuron, the axon can be present or absent in any of the eight lots, except that it must be *somewhere*, that is, it cannot be absent everywhere. This means that there are $2^8 - 1$ (that is, 255) possible axonal configurations. By the same reasoning, we can reach the identical conclusion for any neuron's dendritic tree. Since we can combine any hypothetical axon with any dendrite, there are 255×255 pairs describing the presence or absence of a neuron's axons and dendrites in these eight regions.

The regions invaded by the axonal and dendritic trees of a given neuron determine the connections it can form with other neurons. It is thus reasonable to assume that neurons of the same type must invade the same regions. Therefore, just a binary partition of the brain in each of its three physical dimensions results in more than 65,000 different types of neurons distinguished on the basis of their potential connectivity!

Of course the brain is divided into many more than eight separate regions. In fact the number of distinct brain regions is currently debated among researchers. However, at least in terms of known boundaries separating regions of selectivity for axonal and dendritic penetration of different neurons, we can venture an estimate of 10,000 brain parcels. Yet for the sake of this argument it wouldn't matter much if this number were only 1000 or even 100. The reasoning of the previous paragraph would force us to conclude that any division into twenty or more regions is already sufficient to define a larger number of theoretically possible *neuron types* than the actual count of *physical neurons* in a human brain (a clearly absurd proposition)! Thus, the number of neuron types that *could* exist given the principles of axonal and dendritic connectivity is blatantly astronomical.

The above analytical approach doesn't hold water on rigorous scrutiny, as dendrites do not travel long distances (for reasons discussed in section 3.4). Moreover, if all axons invaded all parcels, the network would run out of available space without any chance to accommodate the necessary number of neurons. Nevertheless, the numerical conclusions are sufficiently overwhelming as to remain unaltered on more careful analyses: only a very small fraction of neuron types can possibly exist out of the imaginable combination of arbor shapes. On the one hand this is still only part of the story in the sense that each of the existing neuron types still must be further characterized by its firing patterns, developmental origin, molecular content, and so forth. On the other hand these arguments strongly suggest that the identification of neurons in types by their axonal and dendritic arbors is core to network formation. Because information processing in the brain is fundamentally a matter of circuitry, we can truly say that the trees of the brain are the roots of the mind.

In an attempt to disentangle the conceptual web of neuron types and their number, let's return to the distinction of protagonists and supporters discussed in section 7.4. The job of protagonist neurons is to process and transmit information content. The job of supporter neurons is to allow and assist protagonists to do theirs by balancing, coordinating, and synchronizing their signals. Recall the evidence pointing to the approximate identification (at least in the cortex) of protagonists with abundant excitatory *principal neurons* projecting long axons. Conversely, supporters correspond to sparser inhibitory *interneurons* with locally confined axons (see table 7.1 at the end of the last chapter).

The relative abundance of principal neurons with respect to interneurons is worth noticing, especially considered together with the greater apparent diversity of interneurons. After all, in Hollywood movie sets, every real protagonist actor is typically accompanied by a large number of assistants, stunts or doubles, and walk-ons; in the credits at the end of the movie, for a cast of twenty recognizable names, the support staff is easily counted in the high hundreds. Similarly, for a couple of physicians in a standard medical office one could find a dozen assistants including nurses, paramedics, interns, and secretaries. Why are there more *types* of interneurons although principal neurons outnumber interneurons?

The puzzle, however, might be only apparent, after all. For a more fitting metaphor, we could compare interneurons to the kind of personnel taking care of requirements that are complementary and conceptually unrelated to the main line of business. Consider for example the cleaning crew of a medical office, the accountant filing the tax return for the practice, and the safety inspector checking that the instruments are up to code. Although the set of professionals required for keeping that medical practice going is highly diverse, each of them performs the same functions for multiple offices. Indeed the diversity of interneurons reflects the variety of roles they perform, but each interneuron works in parallel with several principal neurons, which explains the ratio of their relative abundance. Interestingly, much as an accountant might specialize in physician tax returns, it is reasonable that interneuron types would also diversify depending on the principal neurons they serve.

Where does this leave us in terms of number of neuron types? From the bits of knowledge about the most studied brain areas (notably neocortex, hippocampus, and cerebellum), we can roughly guess that, although in absolute terms principal neurons outnumber interneurons by a factor between five and ten, there are probably between five and ten interneuron types for every type of principal neuron. Thus, the residual question becomes: How many principal neurons are there in a human brain? Eventually, we would like to ascribe a distinct computational function to each type. Because principal neurons are long-projecting, each type could be expected to have a unique connectivity pattern across brain regions (ideally explaining its function). Here the issue becomes tricky because of the possibly arbitrary choice of granularity, not only in terms of brain regions (as discussed above) but also of cognitive function.

Consider, for example, the division of the neocortex into its main functional areas such as visual, auditory, sensory, motor, language production and comprehension, recognition of faces, and so on. By histological examination of soma staining (without accounting for axonal and dendritic arbors), over one hundred years ago Brodmann published a cortical partition into more than fifty parcels, each of which has since been associated with specific cognitive function. Ongoing efforts aim to map the entire cortical connectivity among all Brodmann areas.[9] It stands to reason that the principal neurons of each of these areas should constitute distinct types because they perform different functions and likely correspond to distinguishable connectivity patterns. This resolution, however, is far too coarse to account for the human mind.

The first three Brodmann areas, for instance, correspond to the somatosensory cortex, responding to touch and bodily sensation. This region is organized topographically, meaning that adjacent positions in the brain code for nearby locations on the body. For example, touching the nose and the upper lip would activate neurons with somas located fairly close to each other, whereas touching the nose and a knee would activate more distant neurons within this part of the brain. The resulting map, often referred to as the *homunculus* (Latin for "little person" because it fits in the head!), is organized contralaterally, with the right foot represented in the left cortex and vice versa (figure 8.6). An analogous mapping also exists in the motor cortex, encoding for execution of action commands by various body parts.

We can clearly differentiate touch sensations in each finger and in various positions of each finger (as well as on the tongue, neck, etc.). Thus, it is tempting to assume the existence of as many types of principal neurons in the somatosensory cortex as body spots we can distinguish when touched. At the finest level of granularity, however, this approach might break down because neurons adopt a highly effective computational strategy named *population coding*. This means that a given body location is represented not by a single neuron but by a large number of surrounding neurons each firing at different rates. At first, such an organization might seem to require *more* neurons, not fewer, to distinguish the same number of positions. To understand why this is not the case, imagine the simplified situation of identifying the location on a simple line between two fixed positions (for example, representation of touch on the internal side of your right forearm somewhere between wrist and elbow). Population coding is so efficient that it could represent this whole stretch of body with only two neurons! How could that be possible?

Let's call these two hypothetical principal neurons responding to touch on the forearm the *wrist* and *elbow* cells. Touching the forearm just next to the wrist would activate the *wrist* cell, and tapping the elbow would activate the *elbow* cell. However, a stroke exactly midway between the wrist and the elbow would trigger both neurons to fire at the same rate. A touch a third of the way closer to the wrist would also activate both neurons, but the *elbow* cell would fire at half the rate of its counterpart. Using this clever

Figure 8.6
Homunculus (Daniel Segrè, 2014, Fresh Pond Park, Cambridge, Massachusetts).

tactic, the *wrist* and *elbow* cells could manage to inform the rest of the brain of any part of the forearm being touched!

We can even compute how accurately this encoding could be: it takes approximately one-tenth of a second to detect touch, as the average human reaction time is just around 200 milliseconds, which must account for motor command as well. Each spike lasts about 1 millisecond followed by another millisecond of forced silence (the refractory period, explained in section 3.1). Thus, the maximum number of spikes available is fifty for the *wrist* and *elbow* cells alike, and the minimum is, well, zero. Thus, the imaginary wrist and elbow cells would be able to code for fifty distinct forearm locations (50/0, 49/1, ..., 26/24, 25/25, 24/26, ..., 1/49, and 0/50), which is already beyond the ability of most people!

In reality it's highly implausible that there would only be two types of principal neurons representing the forearm in the somatosensory cortex. Nevertheless, this example provides a powerful explanation as to why population coding defies a direct attempt to count neuron types by cognitive mapping. Similar schemes apply to the motor homunculus and other cortical maps (visual field locations in the visual cortex, sound tones in the auditory cortex, etc.).

Moreover, the efficiency of population coding allows nervous systems to allocate multiple neurons for encoding of the same information. Rather than wasting resources, such redundant representation endows the brain with considerable robustness: the loss or malfunctioning of a few neurons only causes minimal deterioration of signal. In practice this redundancy is not obtained by literally reproducing identical (or randomly varying) neurons within a given position of the cortical map. Instead, neurons tend to specialize along a continuum of representations, such that each cell would display preferential activity at "its" location of expertise. The entire set of neurons would nevertheless fire (each at different rates) to encode any one location. Such a smooth distribution of function may be reflected by a gradual change in connectivity, whereas each neuron might display a probabilistic preference, rather than an absolute specificity, in its input and output connectivity patterns.

Continuous differentiation of neurons might present an insurmountable obstacle to determining the number of neuron types. At the same time, within the subregion of the somatosensory cortex corresponding to each body part, different types of principal neurons are assigned to representing distinct sensations, such as light touch, strong blow, pinch, pain, heat and cold, and so forth. Similarly, each subregion of the visual cortex coding for a given visual field position contains distinct types of principal neurons encoding for various aspects of vision such as color (with different cells for blue, red, and yellow) and contrast (dark signal against bright background or vice versa). These functionally distinct neuron types are not uniformly distributed in their cortical region but rather form local "patches." One possibility to estimate the number of neuron types is to count the number of such patches. A back-of-the-envelope estimation from this presumption yields a figure ranging between 100,000 and 1 million neuron types.

A complementary challenge to determining the number of neuron types is to assign the right number of neurons to each type. A tempting solution is to divide all neurons equally: if there are 100 billion neurons and 1 million neuron types, one could imagine 100,000 neurons in each type. However, this solution is squarely wrong. As we mentioned above, principal neurons are five to ten times more abundant than interneurons, and interneurons are five to ten times more diverse. Thus, each type of principal neuron should be about fifty times more populous than any representative type of interneuron. In practice the situation is likely even more extreme, as principal neurons and interneurons are not evenly distributed among types within their respective classes. It is not unreasonable to hypothesize more than a million neurons in the most abundant neuron types and less than a thousand neurons in the sparsest types.

Estimating these numbers is not just an academic exercise or a matter of curiosity but rather a necessary step in understanding how the brain achieves its cognitive function. You might remember, for example, that dentate gyrus granule cells in the hippocampus act to disentangle cortical representations. Such function is strongly suggested by their very large number relative to the presynaptic source (entorhinal cortex stellate cells) and postsynaptic target (CA3 pyramidal neurons), implying divergence of input and convergence of output. Similarly, there are only a remarkably small number of dentate gyrus chandelier cells, but each of them extends a wide axonal arbor with thousands of terminal candlesticks. As a result, each chandelier cell contacts (and inhibits) a huge number of granule cells at once. The relative abundance ratio between these two neuron types is exquisitely optimized to achieve precisely synchronous coordination of granule cell firing within the fast rhythms of the entire hippocampal network.

More generally, consider a typical assembly kit for a household appliance. Often the instruction manual starts with a "parts list" and an "exploded assembly diagram" to show how the parts go together. An absolutely crucial piece of information is the number of items contained (and presumably needed) for each part type. For example, the parts list might specify that there are fourteen flat washers and twelve hex nuts. These quantities must correspond exactly to the number of times and places the manual will instruct you to assemble the respective parts within the object being built. I'll share a personal story to explain how important this detail is.

After only two weeks in college I was the victim of quite a nasty prank. My loving roommates completely disassembled my entire furniture collection soon after I had finished putting it together from inexpensive assembly kits. Then they mixed up all parts and piled them together in the middle of the room. The totally evil touch was to add (not remove!) a handful of screws and bolts grabbed from a random toolbox. Those extra pieces made it especially challenging to reassemble the bed, desk, drawers, lamp, closet, dresser, blinds, and chairs in their original configurations. The task at hand in this minor trauma can still be considered trivial relative to the triumphant assembly of one of the world's most recognizable aircraft, the Boeing 747 (the double-decked plane nicknamed "Jumbo Jet"). Its more than 6 million parts maintained for over three decades the record for the most complex machine ever built.[10] Would you trust flying in it if the engineers had mistakenly mixed up 50,000 extra fasteners during construction (why, less than 1%)?

Now reconsider the majestic design of a human brain, with its 100 billion parts. If neuroscientists are ever to understand its operating principles at the detailed level of any sort of exploded assembly diagram, you can rest assured that we'll need fairly accurate numerical estimates for each entry in its part list.

9 Brain Branching and the Universe

9.1 The Liquid Jungle

Our exploration of the diversity of neural arbors in the brain revealed many distinguishing features that we know are somehow relevant to neuronal function. For example, the space invaded by axons and dendrites (their location) is intimately related to network connectivity; the actual shape of these trees affects the intrinsic computational properties of both individual neurons and the circuits they operate within; the neurotransmitter released by each neuron altogether determines the effect on all of its targets, and so on. Nevertheless, we also should realistically recognize that the current state of scientific research only offers a vague understanding of many core issues of neural function. We continue to lack a cohesive theory of how large numbers of neurons work together to generate inner feeling and consciousness.

Even our incomplete comprehension of neuronal structure and function has allowed us nonetheless to propose three core principles for linking the nervous system with the mind. We hypothesized in the first principle an identification of mental states and patterns of electric activity in the brain. A consequence of this equivalence is that knowledge, the ability to instantiate a given mental state, is encoded in the connectivity of the network because electric activity in nervous systems flows through connections among neurons. This consideration led to formulation of a structural mechanism for what it means to acquire new knowledge, that is, *to learn something*. Specifically, in the second principle we equated learning to a change in network connectivity through the creation and removal of synapses. This is a distinct process from the common interpretation of synaptic plasticity, meant as the alteration of the strength (rather than existence) of

Figure 9.1
The far-reaching implication of axonal-dendritic overlap. (Top) Non–fast-spiking descending basket cell from layer 2/3 of the rat motor cortex (see note 21 in chapter 3). Dendrites are colored brown, axon green. The inset zooms in on the apical arborization near the soma (rendering by Namra Ansari from the author's lab). (Middle) Layer 4 interneuron from the mouse somatosensory cortex.[1] Soma and dendrites colored yellow, axon purple. The inset zooms in on the arborization near the soma (rendering by Uzma Javed from the author's lab). (Bottom) Somatostatin-expressing interneuron (see note 4 in chapter 2) from the mouse neocortex. Soma and dendrites are colored green, axon pink. The inset zooms in on the arborization near the soma (rendering by Namra Ansari from the author's lab). All reconstructions are freely available online at NeuroMorpho.org; branch thickness was increased to enhance contrast.

synapses. In our framework, synaptic strength determines the ease, rather than the absolute possibility, of instantiating a mental state.

The third principle, which merely builds on the logic of the first two, is nevertheless the most radical and novel proposition of this book: that the spatial proximity of axons and dendrites, enabling synapse formation and elimination, corresponds to the capability for learning. Such a correspondence has far-reaching implications because it directly ties the branching structure of neurons to the fine line separating nurture from nature (figure 9.1). Without the constraint of physical proximity between axons and dendrites, we could be able to learn anything from experience. With this rule in place, in contrast, each of us learns only those aspects of experience that are somehow compatible with our existing knowledge. If on the surface the third principle appears to curb our potential as thinking machines, it in fact endows each individual with an inimitable cognitive identity. This intrinsic cognitive aptitude in life constitutes a deep signature of everyone's intellectual personality. What a dull world would it be if we could all learn exactly the same details, and the acquisition of new knowledge were entirely determined by present experience, without regard for one's personal history!

In order to appreciate how the three principles of the brain-mind relationship work together to make us who we are, let's consider the kinds of neural events that might go on in our brain in the course of a particular series of events. While you enjoy your breakfast coffee still in your PJs after a restful night of sleep, you notice the noise of rain coming through the

Brain Branching and the Universe

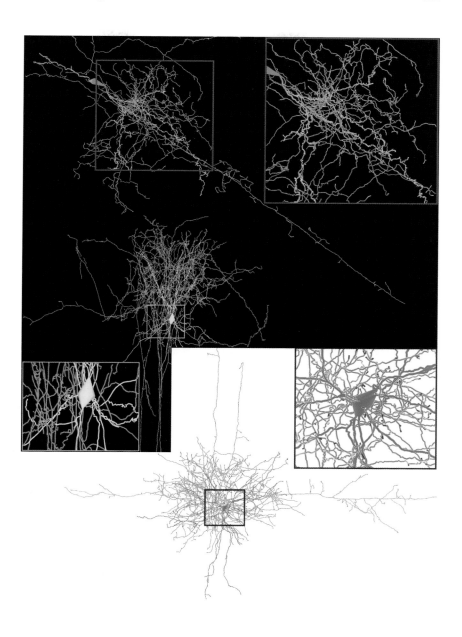

window. You mentally smile at your previous night's wishful plan to bike to work, and you make a mental note to put the umbrella in the car if it's not there already. An hour later as you pull into the office parking lot, you realize you hadn't checked to see if the umbrella was in the car and indeed proceed to make a goofy run for the office building entrance holding an old newspaper over your head. A colleague of yours, who had just finished shaking and closing her umbrella in the lobby, sees you approaching under the rain and kindly comes back out toward you while reopening the umbrella.

What happened to your neurons during this fairly ordinary morning? Let's rewind to the first scene in your kitchen, when your attention shifts from sipping java to the pattering sound of rain. Although you didn't expect bad weather that day, you have heard rain tapping against the window glass many times before. A set of neurons in your brain is already wired up in a circuit suitable to enable the activation of an electric spike pattern underlying the thought of the sound of rain. Indeed, according to the first principle of the brain-mind relationship, the occurrence of your neural activity corresponds to your experience. The awareness of the rain, however, triggers the retrieval of an intention you had conceived of while going to bed the night before (that is, to bike to work), which implicitly assumed clear skies. This means that, unlike most of the fleeting mental states you had before falling asleep, the idea of biking to work had been consolidated into a "prospective memory" (the memory of an imagined future event). Thus, based on the second principle, the neural pattern representing the intended bike ride resulted in the removal and/or creation of appropriate synapses to establish a circuit suitable for reinstating that content at a later time.

Hearing the rain also prompts another neural pattern, one corresponding to the need to put the umbrella in the car, which you later failed to do. Such absence of follow-up action, however, does not imply failure of memory storage (i.e., learning) but rather of retrieval. With appropriate cues, such as seeing the umbrella in the shoe closet, you'd have likely remembered, which suggests that the right synaptic circuit had been formed. As a matter of fact, you will recall that same memory later in the office parking lot! Yet the strength of those synapses, determining the probability of activation, was not sufficient to elicit that pattern while you were leaving the house.

Seeing your colleague coming back out toward you with her umbrella also results in several changes in your brain in addition to reflecting the neural activation pattern directly correlated to that experience. First, you're likely to store this event as an episodic memory, implicating appropriate synapse rearrangement. Moreover, the synaptic circuit encoding the notion of a pleasant and friendly work environment, even though it is already in place, gets reinforced, increasing the chance of instantiating that thought at some other time. You also learn that your colleague's umbrella is red with white stripes, but not that she is wearing pearl earrings, in spite of both images hitting your retina at the same time.

The reason for such mnemonic discrimination lies in the third principle: the umbrella details are represented by neurons that likely have numerous axonal-dendritic overlaps in place with the neurons representing the rest of the episode (e.g., rain, wet ground, running with a newspaper over your head). Those axonal-dendritic overlaps already existed in your brain prior to that day because the respective neurons are connected to nearby neurons encoding for similar or related memories—different umbrellas or locations—but otherwise analogous circumstances.

In contrast, your colleague's pearl earrings fail to become a long-term memory of yours despite constituting an equally visible aspect of the same episode. Although the neurons representing the shiny spheres are coactive with those coding for the rest of the subjective mental state and surrounding scenery, they do not form new synapses with the circuit consolidating that episodic memory ("firing together," but *not* "wiring together"!). Again, the third principle suggests the simple explanation that the axons and dendrites of the respective neurons do not pass near each other within your brain. This fact is itself due to your lifelong past experience, according to which jewelry seldom co-occurs with other content of rainy scenes. Thus, pearls (like rubies and diamonds) would likely be represented somewhere else in the brain where the placement of the underlying neurons ensures minimal waste (i.e., optimal usage) of axonal cable for adequate connectivity.

The imaginary episode of the rainy morning is typical in terms of cognitive dynamics, neural activity, and circuit remodeling in the nervous system. Of course the experience of each person is different every day, and the specific details by which neurons and synapses are activated or changed are indeed unique. Nevertheless, the above *overall* account could be adapted with minimal alterations to describe pretty much any other normal sequence of events. Inner cognitive life corresponds to a continuous and parallel process of perceiving, thinking, and sometimes remembering (and some other times not) past stored information. At the same time, we both adapt the probability to retrieve available knowledge and learn anew, that is, store memories and acquire the ability to access them later.

The basic principles of the brain-mind relationship proposed in this book provide simple neural correlates of the above cognitive operations: perceiving, remembering, and all mental states in general consist of spatial-temporal spiking patterns in brain circuits. Adapting the probability to instantiate one or another mental state corresponds to modifications of synaptic strengths in those circuits. And last, acquiring new memories involves the formation and elimination of synapses.

Section 4.2 describes the gigantic neural network of the brain as a system of roads and the spikes traversing the circuit as traveling vehicles. The roads at once enable but also constrain the traffic patterns, in the sense that they allow cars to go from one place to another but also dictate where the cars cannot go (i.e., along trajectories not corresponding to roadways). This metaphor is intuitively satisfying, but it only tells half of the story. Every spike reaching a synapse not only triggers a postsynaptic response in the next neuron across the cleft but also either strengthens or weakens that connection. Correspondingly, every mental state helps adjust the accessibility of other mental states. Such a mechanism could be imagined akin to a very smart system of road lights continuously fine-tuning the timing of the red-green cycle as each car passes in order to optimize the overall transit time of all drivers. But the brain is even more plastic than that because in addition to adjusting synaptic strengths with every spike, new synapses frequently come and go!

Structural plasticity may be better visualized in terms of a fluvial depiction rather than an urban commute. River beds guide the path of the water, but the water flow at the same time continuously molds the river beds. Similarly, neural connections define the path of electric signals, but those same signals alter the connectivity in return. Of course the time scale of synapse formation and pruning (tens of seconds) is much slower than that of synaptic transmission and spike initiation (tens of milliseconds). However, a spike reaches tens of thousands of synapses and triggers another spike in only a small percentage of the connected neurons. Even if just a small fraction of synapses gets rewired at any one time, when multiplied by the enormous number of synapses in the brain, this tiny proportion still amounts to a substantial recharting of the neural riverbed. Unlike a river sluggishly altering its path over the course of centuries and millennia, the circuit of the brain undergoes substantial recarving on a daily basis. Thus, the relationship between brain structure (network connectivity) and activity (spatial-temporal spiking patterns) is much more like a bidirectional interaction than a one-way causal relationship. The actual flow of information and the *potential* for information flow are mutually affecting each other in a dynamic interplay (refer back to figure 6.4).

9.2 The Uniqueness of the Individual

Mapping of the human genome revealed that two unrelated individuals of the same gender and ethnicity share more than 99.9% of their genetic sequence. The remaining 0.1% accounts for traits associated with individual diversity such as eye color, body shape (e.g., height), predisposition to certain diseases, and so forth. It might initially come as a surprise that only one in one thousand genetically determined aspects of human identity differ between two subjects. Such an estimate, however, makes sense when one considers all basic infrastructures that genetic information must encode: we all have two arms, each ending with a hand and five distinct fingers in the proper order. Likewise, we all have a liver, a heart, two kidneys, and two lungs. It is easy to see that the list becomes so long so quickly that the physical differences between any two individuals (let's call them Lisa and Monica) are indeed a very small minority. Lisa has blue eyes, is tall, and has a predisposition for breast tumors, whereas Monica has brown eyes, is short, and has no family history of cancer.

Yet Lisa and Monica are profoundly different people. They have different lifestyles, like different music, and have completely opposite personal talents. What's more, it goes almost without saying, they have different childhood memories, different plans for the future, and different factual knowledge. Lisa lives downtown, by herself, and is the CEO of a multinational company; Monica lives in the suburbs as a "soccer mom" with three young children. Oh and I forgot to mention, they speak different languages and would not even understand each other if they ever met. If one were to put Lisa and Monica in similar situations, they would understand different details, recall different memories, make different decisions, and learn different associations. For example, Monica would be lost in a meeting of the board of trustees in Lisa's corporation, and Lisa would be a fish out of water going through the school backpacks of Monica's kids. *Even though their genes and body organizations are 99.9% similar, Lisa's and Monica's inner cognitive lives and conscious experiences seem to be 99.9% different.*

What makes Lisa, Lisa and Monica, Monica? *What makes me, me and you, you?* Lisa's and Monica's mental states correspond, says the first principle, to spatial-temporal patterns of neuronal activity in their brains. Thus, the complete difference of the mental states they experience in similar context indicates that their respective neural activity patterns have almost nothing to do with each other even in response to identical stimuli, such as the sight of one and the same kid's backpack content. In this (or equivalent) imagined scenario of Lisa and Monica both staring at the same open backpack from the same distance and visual perspective, the images hitting their respective retinas are the same. Thus, the patterns of spikes traveling down their optic nerves would likely be almost identical. In fact, Lisa and Monica *see* the same homework sheets and read the same letters, but their cognitive and emotional reactions are nothing alike. Monica is proud of her child's progress in math and realizes that she should go talk to the history teacher, all the while noticing that the lunch box is missing from the backpack. The very same visual scene appears to Lisa just as a typical grade school backpack with no particular meaning.

The patterns of spikes in Lisa's and Monica's brains remain similar beyond their optic nerves and through the parts of their nervous system recognizing letters (math and history homework) and objects (backpack and lunchbox). Nevertheless, within a small fraction of a second this same initial neural activity triggers dissimilar spike patterns in the rest of Lisa's and Monica's brains. This is because their networks are *wired* differently. Monica *knows* her son's school assignments and the look of his lunchbox; Lisa doesn't. If you recall from section 4.3, this means that some neurons in Monica's brain are interconnected to form circuitry that, when activated, recalls that particular mental content. Lisa's brain lacks that specific connectivity—but of course *her* network is structured so as to recognize the faces of her corporation's trustees (and the subtlest meanings of their facial expressions). The uniquely different mental states of two individuals reflect the extensive differences of neuronal connectivity in their brains.[2]

Clearly, Monica's neural circuitry encoding knowledge of her kid's schooling and Lisa's neural circuitry encoding knowledge of her board of trustees were not determined genetically but by their prior experience. The second principle explains that those synapses formed as Monica and Lisa learned, earlier in their lives, the corresponding associations, such as the handwriting of the math teacher or the insincere smile of the senior trustee. Much of one's individual knowledge, of his or her thousand trillions of contacts between axons and dendrites, are the result of the whole life history of that person, second by second and year by year. Our mental states are so different (even in reaction to the same context and stimuli) because our network connectivity is vastly dissimilar. Every person's brain circuitry is unique as a result of the exquisitely distinct sequence of events each of us experienced throughout our conscious existence.

The flip side is that every mental state alters the circuitry in return, and as a consequence, we *keep learning*. Monica will remember the following day to congratulate her child on his school success, to email the teacher for an appointment, and to call the lost-and-found office inquiring about the lunchbox. Lisa will write a detailed report of the board meeting recalling several details of that day's events from her memory.

One of the most remarkable features distinguishing Monica and Lisa (and any two individuals) is not just the history of their lives and what they *know* but rather what they can and cannot *learn* from present and future experience. If Monica had attended the board of trustee meeting of Lisa's company, she would not later be able to remember nearly any relevant details to include in a corporate report. Conversely, only a few hours after briefly inspecting the backpack content, Lisa would have already forgotten the homework grade, let alone the idea of emailing the teacher. And we haven't even considered the fact that these two fictional characters are supposed to speak different languages! Though Monica would have heard the same utterances at the board meeting, she would not have understood a word.

The third principle provides a straight neural correlate of each person's acutely distinct learning ability: neurons can form new contacts (and thus learning can occur) only if the axonal tree of the sender and the dendritic tree of the receiver have overlapping branches. By determining the potential for establishing new synapses, the shape of these arbors thus controls the selective acquisition of new knowledge. In this sense these trees of the brains effectively constitute the determinants of the mind.

Every person lives through the unique flow of physical events his or her body is embedded within. Both the mind and the brain of each individual continuously adapt to such exclusive perspective. At any one instant a person's mental state corresponds to the pattern of neuronal activity in his or her brain, which in turn depends on the surrounding physical events and the brain's network connectivity. The occurrence of that activity pattern and the experience of the corresponding mental state trigger learning by modifying the synaptic circuitry, but this occurs only if the underlying neuronal trees are suitably shaped and positioned. This "instantaneous learning potential" is itself molded by experience (refer back to figure 6.4).

Although the mental states, experience, and learning potential of each person are indeed unique, it is nonetheless possible to recognize and consider some common aspects across individuals. We all can experience the same or at least similar emotions, such as happiness, anger, awe, surprise, fear, satisfaction, and curiosity, to name just a few. We can all relate to particular situations, such as the joyful wedding festivity of a dear friend, the frustration of being stuck in traffic while late for a meeting, the "a-ha!" moment of grasping a difficult concept, the embarrassment for an awkward tongue slip during a public speech, or the irresistible laughter after a funny joke. And we can also all imagine emotionally neutral scenes, such as walking in a grocery store while pushing a shopping cart and whistling a tune from our childhood.

In each of the above examples the *details* of the mental state would be different for you, me, Lisa, and Monica. The specific *meeting* each of us would be heading to when stuck in traffic would be different, and so would be the *groceries* in our respective carts, or the whistled *tunes*. Yet my mental state while sitting in traffic would be more similar to that experienced by Lisa in the same situation than, say, that experienced by either of us in the grocery store. There must be some aspects of brain connectivity that are common among (most) human beings. The limited resolution of current technology only allows us to identify brain areas generally responsible to perform basic visual tasks (such as detecting an edge or color contrast), recognizing human faces or common tools, planning a spatial path, or repeating a sequence of words. This is too coarse a measurement to investigate the maximum common denominator between the minds of two people.

Still, if the connectivity of a human being is the substrate for his or her knowledge, if the activity patterns constitute the neural correlates of his or her mental states, and if the axonal-dendritic overlaps provide a blueprint of his or her learning potential, we can at least imagine the equivalent correspondences across all of humanity. In order to conceive of characterizing the entire synaptic connectivity of the human brain, we could rely on population-level statistics and leverage the notion of neuron types. Instead of a synapse either being present or not between two neurons, we could measure a probability of that synapse being there between neurons of the corresponding types. For example, a person either knows or doesn't know that brain cells are shaped like trees. Monica might have that synapse in place, Lisa might not, and if we take a representative sample of human beings we might venture to estimate that 78.6% of the population is missing that neuronal contact and interesting piece of corresponding information.

The above scenario is clearly oversimplified, not only because knowledge of facts does not correspond to individual synapses but also because this hypothetical thought experiment presumes that we can identify a functionally equivalent pair of neurons (and thus determine if they are connected) from one person to the next. This is not typically the case, but there are sophisticated mathematical and statistical approaches to circumvent this difficulty. Moreover, we could think of asking the same questions regarding axonal-dendritic overlaps and learning potential instead of synapses and knowledge or the presence of a (set of) spike(s) and of a corresponding mental state among individuals. At least with a bit of imagination it is possible to envision an "average brain" or rather a "probabilistic brain" representing all human beings, their commonalities, as well as the typical intersubject variation.[3]

We still lack sufficient data to understand the specific correspondence between activity patterns and mental states, between neuronal connectivity and knowledge, and between axonal-dendritic overlap and learning potential. We understand enough about the functioning of the nervous system, however, to know that measuring these properties will be important to characterize the precise relationships between brain structure and mental dynamics. Pinning down this information for not just one but many individuals will lead to a deep comprehension of how the brain gives rise to the mind, both in exact terms for a single person and more generally as the cognitive meaning of humanity.

9.3 Brain Scans and Super-duper Microscopes

It is common to read in the news these days of brain-imaging studies reporting on the neural correlates of gambling and trust, on the prospect to build reliable lie detectors based on brain signals, on the potential for noninvasive neuroimaging to diagnose autism or schizophrenia, or even on flat-out mind reading based on neural recording. Indeed, there are several techniques capable of peering into the human brain noninvasively, which is to say from the outside across the skull of a normal, awake individual. One relatively simple method is called electroencephalography or EEG. In EEG, a number of electrodes (probes designed to detect electric currents) are positioned on the surface of a person's head and relay the measured signals to a recording computer via wired or wireless transmission. Because neurons encode and exchange information via electric currents, the signals picked up by the electrodes provide a measure of brain activity.

The excellent temporal resolution of EEG allows discrimination of signals at or even below the level of milliseconds and would thus in principle be suitable to identify individual spikes and synaptic potentials. The spatial resolution, however, is limited by the restricted number of electrodes that can be fitted on the head. Even so-called high-density EEG has no more than 256 electrodes, and any signal is thus ascribed to a grand average of neuronal activity within 1 of only 256 brain regions. Comparing this number to the ~10^{11} neurons of the human brain leaves as many as 300 million neurons under each electrode! Another severe limitation of EEG is that it only picks up activity from the brain surface because deeper currents are shielded by the electric signals above. Even the superficial signals are filtered and distorted by the skull, skin, and hair.

Another popular brain-scanning technique is called functional magnetic resonance imaging or fMRI. This approach exploits a particular physical property of hemoglobin, the blood oxygen carrier: when hemoglobin binds oxygen, it loses its magnetic property. Recall that after a neuron fires an action potential, its molecular pump must "recharge" opposite gradients of sodium and potassium across the neural membrane. This process is energetically demanding and thus temporarily depletes oxygen in the nearby blood capillaries. Within a couple of seconds oxygenated blood flows to the site of activity to restore the baseline oxygen concentration. fMRI measures the local transient increase in magnetism that accompanies neuronal activation in a given brain region. Relative to EEG, fMRI boasts a superior spatial resolution, as it can localize neural activity down to within a cubic millimeter of volume. However, its temporal resolution is of the order of one or more seconds and thus far poorer than EEG or what would be relevant for single neural events. Even fMRI only detects the overall activation of hundreds of thousands of neurons rather than individual neurons.

Measurement of neural activation via EEG or fMRI can be analyzed to infer functional connectivity among the activated regions. Essentially, if a region systematically activates another region, this may be due to a connection between the two, and appropriate statistical analysis can compute the most likely connectivity of a network given the evidence provided by the entire patterns of activation over time. Another brain-scanning technique, called diffusion tensor imaging (DTI) or diffusion MRI, leverages the magnetic property of the hydrogen atoms contained in water molecules to track fiber tracts. Unlike fMRI, DTI measures structure, not activity. Specifically, it visualizes large collections of axons traveling together from one region of the brain to another. Section 1.2 mentioned the thick axonal bundles called the corpus callosum connecting the right and left cortical hemispheres. This and many other such "highway systems" in the brain are amenable to DTI visualization. Because these axonal tracts originate from groups of neurons to reach distant dendritic targets, DTI depicts the main connectivity pathways among brain regions. Nevertheless, DTI cannot indicate the directionality of the connection (which region is the sender and which is the receiver). At any rate, as for fMRI, DTI lacks the spatial resolution necessary to come even remotely close to an individual neuron arborization.

Despite the limitations of EEG, fMRI, and DTI, the ability to measure and somewhat localize activity, dynamics, and structure in the brain noninvasively in awake, cognizant human beings has already enabled tremendous progress in relating mental states to neural substrates. Much of what is known about the role of various cortical areas, such as those involved in language comprehension or production, visual and other sensory modalities, facial or object recognition, motor control, logical thinking, and so on is nowadays derived or validated by the above brain-imaging techniques or one of their many variants.

Even the strongest correlation between activity in one area and a corresponding mental state or behavior does not necessarily prove causality. For example, the systematic co-occurrence of the activation of Broca's language production area with word uttering cannot be taken to imply that Broca's area *makes* the subject speak. Logically, there could be another physical event (say, the activation of another brain area) triggering both activity in Broca's area and speaking. This possibility can be excluded by the use of another technique called transcranial magnetic stimulation or TMS, which temporarily derails activity in a target brain region by inducing a strong electric current that interferes with the spontaneous neural activity. If TMS application to Broca's area alters or disrupts syntactic processing (which it does), the involvement of this region in that function is indeed confirmed.

Although useful for many applications, ultimately EEG, fMRI, DTI, and TMS cannot probe the essential foundation of the brain-mind relationship: the activity, connectivity, and plasticity in networks of neurons. To be fair, DTI does map connectivity in the brain, but the elements of this map are brain regions rather than neurons. Similarly, fMRI measures activity, but the unitary players of that activity are again brain regions instead of neurons. Likewise, TMS alters that activity at the level of brain regions. Understanding the deep relationship between brain and mind in terms of the inputs and outputs of the relevant processing substrates, axonal and dendritic trees, requires just the kind of mapping and measurement allowed by DTI, fMRI, and TMS, except not on brain regions but on neurons (table 9.1). The real challenge is to be able to do so not in one, one hundred, one thousand, or one hundred thousand neurons at a time but in the whole brain at once.[4]

Brain researchers have been attempting to map neurons and their networks since the early days of neuroscience. Modern techniques allow unprecedented access to a great deal of information about neuronal structure, activity, and function, at least in animal models. One approach involves the surgical implant of silicon probe arrays into the brain of a rat or mouse. Although the procedure is invasive and cannot be applied to humans, this technique allows recording of every single spike from several

Table 9.1
Spatial and temporal resolution of macroscopic and microscopic neural recording

Technique	Span	Spatial resolution	Temporal resolution
EEG (activity)	Whole brain	Centimeters	Milliseconds
fMRI (activity)	Whole brain	Millimeters	Seconds
DTI (connectivity)	Whole brain	Millimeters	Static (structure only)
TMS (transient lesion)	Whole brain	Centimeters	Seconds
Extracellular recording	Hundreds of neurons	N/A* (activity only)	Milliseconds
Intracellular recording	Single neuron	N/A* (activity only)	Milliseconds
Light microscopy	Centimeters	Micrometers	Static (structure only)
Electron microscopy	Millimeters	Nanometers	Static (structure only)
Voltage sensors	Millimeters	Tens of micrometers	Tens of milliseconds
Optogenetics	Hundreds of neurons	Tens of micrometers	Milliseconds

*N/A, not applicable.

hundred neurons in real time while the rodent is performing typical behavioral tasks such as finding a spatial path, foraging for food, mating, sleeping, and dreaming. An alternative is to insert a single electrode precisely juxtaposed to the soma of an individual neuron, then utilize pressure to create a microscopic hole through the membrane. This technology, named *patch clamp*, enables recording not only of that neuron's spiking pattern but also, with much greater sensitivity, of all the excitatory and inhibitory signals from the thousands of presynaptic neurons reaching that soma from its entire dendritic tree. Detection of synaptic inputs and their variation over time is essential for studying synaptic plasticity in response to experience and activity.

Once a neuron is "patched," it is possible to inject into it a dye through the electrode. The dye diffuses through the dendrites and axons and can thus reveal the structure of the whole arbors by microscopic imaging. Many neurons illustrated in the figures of this book have been visualized precisely in this manner.[5] Intracellular dye injection, however, only works one neuron at a time. One exciting recent development involves the genetic modification of mice or other animals (such as fruit flies) to make their neurons spontaneously produce fluorescent proteins (genetically engineered from jellyfish) of various colors. Examining the nervous systems of these creatures under a light microscope renders a stunning forest of green, blue, red, and yellow neurons. Genetic fluorescent reporters eliminate the need to stain neurons one by one, but they may face the opposite challenge: if *all* neurons were colored, *none* would be visible, because brain space is filled with neuronal trees. Even if every neuron is given a different color (a clever genetic technique called *brainbow*), remember that the volume invaded by the axonal tree of a single projection neuron is shared by the arbors of potentially billions of other neurons. There simply aren't enough different colors to allow distinguishing so many neurons by either the human eye or the most sophisticated color-measuring machines (spectroscopes).

Therefore, no matter how axons and dendrites are rendered visible, optical microscopy ultimately requires that only a small percentage of neurons be visualized. Neuroscientists obtain such sparse labeling by linking the genetic instruction to produce the fluorescent proteins to certain specific genes that are only expressed in some but not all neurons. For example, you may recall from section 3.2 that most inhibitory neurons synthesize and release the neurotransmitter GABA. In contrast, most excitatory neurons release a different neurotransmitter (glutamate) and do not express the gene responsible for synthesizing GABA. If the genetic instructions to produce green fluorescent protein are attached to the gene encoding for GABA synthesis, the resulting animals will have fluorescent green inhibitory neurons and normal excitatory neurons.

These kinds of technological feats thus enable coloring a huge number of neurons in a single brain at once while leaving an even greater number of neurons invisible. And still all of that doesn't tackle the hardest part of the problem, which is not *just* to image the microscopic tree-like structure of neurons but actually to extract these data from the images. Even if a magic pencil could trace axonal branches at the speed of a commercial jetliner (500 miles per hour), it would still take more than a year of drawing to complete a single human brain! Today most of the so-called reconstruction of neuronal geometry is still carried out manually in thousands of neuroscience laboratories worldwide. Clearly, it is necessary to devise fully automated software programs that can perform these tasks algorithmically.[6]

There are additional technical challenges to the ability to visualize the brain's neuronal network. Synapses, as well as the thinnest axonal branches, are just too small to be detected by light. Light travels through space in the form of a wave, and the location of the objects it illuminates can only be determined within the approximation of the distance between two consecutive wave peaks. This distance is of the order of one-half of a millionth of a meter or one-hundredth the thickness of a typical human hair. Although this is undoubtedly very small, axons can be ten times thinner and synapses twice as thin as the thinnest axons!

Such ultrathin structure doesn't mean that axons and synapses are invisible by optical microscopy: we can still genetically encode fluorescent proteins in thin axons or other similar reporters in synapses. Their image will, however, be blurred like a badly out-of-focus picture, and we won't be able to recognize their details. This is unfortunate because almost every synapse in the brain is surrounded by many more axonal and dendritic branches than the two specific branches from which it arises. Thus, even if we could image every neuron and all synapses by light microscopy, it might still be impossible to tell which neuron is connected to which and by what synapses.

The limit of light resolution can be surpassed by adopting a different type of microscopy, which utilizes beans of electrons instead of light. Electron microscopy has a much greater magnification power than light microscopy and can routinely detect not only thin axons and synapses but also their inner structures, including neurotransmitter-containing vesicles in the axonal terminal and the dense clumps of their receptors on the opposite (dendritic) bank of the synapse. Such superior resolving power also allows detection of all the parts of each and every neuron within a given volume because it enables us to see the distinct membranes of two adjacent branches and even the minuscule space in between. Electron microscopy can in principle achieve *dense* reconstruction of the nervous tissue ("no synapse left behind") as opposed to the sparse sampling of gross connectivity discussed above.

Despite its virtues, electron microscopy has three damning vices of its own. First, relative to optical imaging, the data extraction process in electron microscopy is many times slower. In light microscopy branches are traced as lines (with a thickness) meandering in three dimensions. An entire branch can be traced with a single stroke of pen from a view parallel to the axonal trajectory. Instead, electron microscopy detects the outer membrane of the axons, and branches must be traced with a huge number of consecutive contour lines from the front view (like drawing a botanical tree by following the texture of its whole bark). Second, the field of view of electron microscopy is much smaller than that of optical microscopy. Although it is in principle possible to image an entire mouse brain by optical microscopy, the largest volumes captured by electron microscopy to date are one thousand times tinier. The third flaw is literally lethal: electron microscopy contrast agents can only penetrate fixated (dead) tissue. In other words, to see neurons with electron microcopy, you have to kill them first.

Microscopic visualization of neural structure can only reveal a partial story: after all, the main job of neurons is to exchange electrical information with each other. Clever genetic engineering has already begun to break the barrier of static microscopic imaging, bringing spiking dynamics within the reach of optical microscopy. Specifically, imagine a conditional fluorescent protein that only turns green on transient passage of electric current. Such voltage sensors can effectively detect neuronal communication without inserting an electrode in the tissue! Unfortunately, the imaging contrast of voltage-sensitive fluorescent proteins is still insufficient for reliable signal detection other than in the cell body (as opposed to dendrites and axons). Moreover, the temporal resolution is also limited to spike trains rather than individual action potentials.

More and more tricks are coming out of the magic hat of molecular biology, including the tools of *optogenetics*. You may remember two types of ionic channels described in chapters 2 and 3: those gated by voltage and those gated by neurotransmitters. At the same time, section 7.3 mentioned rods and cones, special neurons in the retina endowed with ionic channels gated by light (photoreceptors). Exquisite genetic modifications have allowed the chimeric design of highly selective synthetic photoreceptor proteins that can be inserted in specific neuron types. Pointing a laser beam of appropriate color (say, blue) on these neurons may cause the opening of sodium channels, triggering a pattern of spikes corresponding to the pulse of light. But that's not all: pointing another laser beam of a different color (say, red) on the same neurons could open chloride channels, causing the spiking activity to halt. These optogenetic interventions effectively insert into neurons nanoscale "on" and "off" switches that neuroscientists can control with laser beams (figure 9.2).

Figure 9.2
Peering into the future. (Top) Optogenetics (Daniel Segrè, 2014, Halls Pond Sanctuary, Brookline, Massachusetts). (Bottom) Reaching for the sky (photograph by Daniel Segrè).

The correspondence between neural activity, structure, and plasticity on the one hand, and essential cognitive functions (mental states, knowledge, and learning) on the other that we proposed in chapters 4, 5, and 6 can be revisited now in light of the modern tools of microscopic imaging, genetic engineering, and optogenetics.

According to the first principle of the brain-mind relationship, any thought, emotion, percept, or intention we can feel or imagine corresponds to spatial-temporal activity patterns in the central nervous system. At the level of the whole human brain, these dynamics can be glanced on noninvasively by fMRI and even altered by TMS. However, these approaches are far off in terms of both spatial and temporal scales, resembling improbable attempts to follow the flight of a bumblebee from satellite imagery or to play Rachmaninoff's piano arrangement of the homonymous Rimsky-Korsakov's fast orchestral interlude with a set of construction cranes. At the more relevant level of neurons and action potentials, spiking dynamics can in principle be detected with electrodes or optical recording, although their invasiveness excludes routine human application. Moreover, neither of these techniques scales to large enough brain regions to generally enable mental state mapping.

Learning, states the second principle of the brain-mind relationship, is nothing but formation and elimination of neuronal connections. Correspondingly, the whole knowledge of an individual, that is, the instantaneous ability to access the entire repertoire of mental states each person is capable of, is stored in his or her full connectome, determining which activity patterns are physically possible. Noninvasive brain imaging of human structural and functional connectomics can be obtained with DTI (or diffusion MRI) and analysis of fMRI, respectively. However, as before, the resolution afforded by these methods is still off by several orders of magnitude from the level of synapses.

As we discuss in this chapter, the technique of choice to identify synaptic connections is electron microscopy. Individual functional connections between neuron pairs can be measured with dual intracellular recording. Small circuit connectivity can be gleaned from optical recording with voltage indicators through analyses similar to those used in human brain connectomics by fMRI. But once again, these microscopic technologies are not applicable to live human subjects, and whole brains of even the smallest mammals are still out of reach of their current scope.

The third and most audacious principle of the brain-mind relationship equates the spatial proximity of axons and dendrites from specific neurons to the capability of learning. Although this is certainly the least broadly known (let alone accepted) of the three main principles proposed in this book, perhaps it also constitutes the scientific hypothesis closest to being experimentally testable. As pointed out before, the spatial resolution of human whole-brain imaging is far too coarse to distinguish axonal-dendritic proximity from synapses. Similarly, the temporal scale of noninvasive connectomics is just too limited to catch learning in the act and can only offer static snapshots. Fluorescence optical microscopy, in contrast, has the potential to yield in the not-too-distant future relevant data on axonal-dendritic overlaps for a substantial sample of neurons in an entire mammalian brain, at least in animal models such as mice. Machines have already been designed capable of scanning entire mouse brains at the resolution of dendritic and axonal branches (macroscopes). Moreover, thanks to genetically encoded reporters, this approach is suitable for time-lapse investigation of live behaving animals. Thus, experience-dependent structural plasticity will likely be the first neural correlate of cognition to be mapped systematically.

9.4 Neurofuturism: The Code Is the Beginning, Not the End

What does it *mean* to understand the brain and its relationship with the mind? At the end of section 1.1 we argued that a solution to the mind-brain conundrum will ultimately require a mathematical formulation identifying mental and neural observables. We have direct access to our own mental content, and we can report it to others by verbal and nonverbal communication (more on this below). But to quantify neural states appropriately, we have to define the proper level of observation. In the course of this book we identified neurons and spikes as the most likely candidates and the tree-like structures of their axons and dendrites as the key players in the ensuing computational process. Thus, the technological challenge: Can we measure all neural connections and communications in a brain as they unfold over time? And importantly, can we do it seamlessly enough so as to allow a subject to introspect in the meantime, with the goal of characterizing his or her own mental state as precisely as possible, without distractions?

As we describe in section 9.3, human brain structure and activity can be recorded noninvasively by magnetic resonance imaging and even altered by transcranial magnetic stimulation. The experimental setup is constrained, as the subject has to stay still with his or her head inside a noisy scanner rather than going about a normal life. Most importantly, however, these techniques fail to reveal neurons and their activity. At the opposite end of the spectrum we have optical and electron microscopy, which enable us to investigate neurons directly. These two microscopic techniques have complementary strengths and weaknesses: optical microscopy can leverage genetic encoding of fluorescent proteins and voltage sensors in selected neuronal targets and is compatible with laser-controlled activation and deactivation of neuronal spiking. However, the spatial resolution of light is insufficient to unambiguously and precisely detect the thinnest neuronal branches and synapses. In contrast, electron microscopy can reliably visualize even neuronal membranes, subcellular organelles, and the fine inner structures of axons and dendrites; but alas, it can do this only in defunct organisms.

Luckily, scientific progress never rests, and recently the combined application of sophisticated mathematical and engineering techniques has led to the design of several new variants of fluorescent microscopy. With this new technology, resolution is *not* limited by the wavelength of light. This so-called *superresolution* optical imaging affords a visualization power similar to that of electron microscopy while enabling genetic encoding and applicability in principle to live behaving animals.[7] Yet the technical challenge is not over: even if neurons are colored genetically and imaged by superresolution, in order to visualize their trees deep in the tissue it was until recently necessary to cut thin slices from the brain to enable a beam of light to pass through. Because brain slicing can hardly be considered noninvasive, both in vivo optical recording and in vivo optogenetics have so far been largely limited to regions of the brain close to the surface.

Once again, however, recent breakthroughs are rapidly dissolving these previous limitations. Chemical processes including application of penetrating detergents and molecular substitutions of transparent gels for opaque lipids now enable researchers to render mouse brains almost completely transparent! This is an important advancement because it allows us to visualize genetically colored neurons deep in the tissue without slicing. The activity of these neurons can meanwhile be turned on and off optogenetically (being transparent to light works equally well from the inside out and in the opposite direction).

See-through neural tissue does not provide a complete solution because the mouse brain still has to be taken out of the skull before it can be soaked in a cleansing solution. Yet future developments seem within reach, such as flushing the clarifying chemicals through the bloodstream or administering them to the animal through drinking water. More generally, the inexorable march of scientific advances is becoming clear: by combining various approaches along the lines described above it may soon be possible to record an arbitrarily large number of neurons, connections, and spikes from the entire brain of living mice.

To be sure, some problems remain, foremost among them the scalability of any and all of these techniques to the staggering numbers of a mammalian brain (see section 2.4). Soon-to-be assembled macroscopes should allow imaging of a whole mouse brain with single-neuron resolution, but the expected acquisition time will still be of the order of a month per animal. Unfortunately the animal will move, think, and change during that month, so the data will not tell us about the activity of the neurons, only their static positions.

In order to achieve real-time recording of neuronal spikes, the visualization speed has to be reduced to a couple of milliseconds, a whopping nine orders of magnitude (a billionfold). Extension of this design to superresolution techniques ups the challenge even further given the smaller field of view and slower acquisition time that accompany higher magnification.

Despite these difficulties, the history of scientific discovery gives plenty of reasons for optimism. Only eighty years have passed in physics from the 1932 discovery of the neutron, the last "classic" atomic particle, and the 2012 discovery of the Higgs boson, the last building block of the standard model of quantum mechanics. Similarly, the ratio between speed and cost of DNA sequencing has increased even more dramatically in the past four decades.

Peering into the future of neurotechnology, we find it tempting to ponder the question: Will we ever know it all? The most directly foreseeable research points to a massive-scale statistical understanding of some sort of "average brain" for a model species such as the mouse. In other words we will not necessarily characterize any one individual mouse (let alone every single mouse), but only a detailed cumulative representation of an average mouse brain assembled from sparse sampling of many genetically similar mice. The average is only part of the story: we also need to ascertain how much variance there is from this average. Of course we already know much about interindividual diversity even within genetically uniform animal populations. The average neuron-level mouse brain representation will then be augmented with specific correlations between neuron-level statistical brain traits and observable behavioral differences.

There are hard technical reasons for predicting earlier success in statistical brainwide neuron mapping at the overall species level than on a personalized neuron-by-neuron level in a single brain. The set of techniques currently in use, such as coloring neurons, are by construction limited to sparse sampling. We can't visualize all neurons in a given brain region at once because in that case none would be visible in the packed filled space. We need statistics to extrapolate from these limited samples to the whole population of neurons in the brain.

In theory, there *is* a way to achieve dense mapping with sparse visualization. Recall that neuronal firing is temporally sparse. On average, neurons only fire a handful of spikes per second. Because each spike lasts only a millisecond or two, fewer than 1% of the neurons in the brain are active at any one time. If voltage-sensitive fluorescence proteins could be designed that enabled the quasi-instantaneous visualization of the entire axonal and dendritic arbors, their pervasive genetic insertion in all neurons would enable sparse sampling every millisecond, and within a few minutes the whole brain could be mapped.

This scenario effectively requires whole-brain, single-neuron, *real-time* (that is, millisecond-scale), superresolution voltage detection, a monumental goal quite unlikely to be met for many years. Nevertheless, dense mapping may not be truly necessary to understand how the brain gives rise to cognition. A comprehensive statistical mapping of a whole brain for any mammalian species at the level of individual neuron types might do the trick already.

Such a guidebook would reveal how the connectivity among neurons relates to that among regions, establishing a profound link between micro- and macroscales. This would, for instance, help delineate the distinction between projection neurons (extending axons to other brain regions) and local interneurons (with largely or exclusively contained axonal spans). Accurate numbers and interconnection probabilities for all types of neuron would enable the construction of real-scale computer models of the mammalian brain. The initial states of these models would be specified stochastically (i.e., randomly) within the given statistical distributions.[8] However, because the functional and structural plasticity of each neuron type would also be quantified, these models could be programmed to evolve their synaptic circuit by interacting with a real or virtual environment.[9] These simulations could lead to a deep and broad understanding of the brain mechanisms responsible for behavior and the corresponding cognitive states. In particular, the computational role of all neuron types (protagonists and supporters alike) would be determined and related to their connectivity, excitability, plasticity, and molecular identity.

Comprehensive neuron-level statistical mapping of a species' brain paired with stochastic computational models and real-scale simulations might well circumvent the need for individual identification of every single synapse in any particular brain or even in an average brain. There is, nonetheless, still an unavoidable "gotcha" clouding the prospect of a satisfactory solution to the millennia-old mind-body problem. Even sparse sampling of neuronal statistics involves an experimental approach that cannot quite be considered noninvasive. In the best possible scenario we are talking about massive genetic engineering of neural fluorescent reporters coupled with either extreme chemical treatment to make the brain transparent or neurosurgical implants of fiber optics. Either option is understandably untenable for human investigation. The current path of neuroscience research may soon enable us to fully chart the mind of a mouse. Unraveling the neural substrates of the quintessential cognitive traits of *Homo sapiens*, however, might require development of radically different approaches likely involving nanotechnology.[10] At the very least the ability to understand the mouse mind from exploring its brain would help clearly define the technical and scientific requirements to reproduce the same feat in humans.

Eventually, we can assume that we will gain the ability to measure each spike of every neuron throughout the brain of any human being.[11] What then? Understanding how the brain implements the mind is a deeply rooted mystery in the human soul. Cracking that code has extraordinarily profound implications for religion, philosophy, science, and nature. Notwithstanding anyone's personal belief or philosophical attitude, there is no doubt that the practical consequences of such a scientific achievement would be nothing short of revolutionary.

Perhaps the most immediate results would come from medicine. The broad array of neurological and psychiatric conditions afflicting an increasing proportion of the human population would become amenable to cure and even prevention. The quantitative comprehension of the neural mechanisms of behavior and cognition would likely lead to countless biomedical and psychological interventions in nonpathological states as well. These developments may aim to enhance performance or optimize general quality of life.

A parallel benefit of a neuron-level map of brain structure, dynamics, and plasticity will undoubtedly be in the creation of intelligent machines, namely computers and robots or, more generally, artificial agents. Computers are already much faster, more precise, and more reliable than human brains for any operations that can be expressed in algorithmic form. However, we have so far failed to design machines capable of empathy, humor, or even simple perception tasks such as following one or another line of conversation at a busy (and noisy) cocktail party. Understanding these capabilities in nervous systems should enable their engineering in hardware or software implementations. Computers will learn from and adapt to their users; semiautonomous vehicles will learn from and adapt to their environments. The impact on society will be so profound as to become very quickly unpredictable in its detail, except for the high-level forecast of pervasive and exhilarating alterations at all levels of our lives.

Despite the unquestionably immense impact of biomedical and engineering applications, I suspect that the most radical consequences of decoding the mind in terms of neural computation will be felt on the direct interpersonal relationships between individuals.

Until today, and for the entire history of their existence on planet Earth, human beings have always been limited in their ability to communicate inner mental states. All we have is language—not necessarily just verbal language, but also body language, artistic expression (music, paintings, even architectural design), and scientific exchanges (a mathematical theory expressed in formulas, for instance). But all of these linguistic forms of information exchange remain fundamentally indirect relative to actual experience itself. Of course not all feelings, thoughts, emotions, memories, fears, plans, and dreams are *meant* for communication. Most mental states are intended solely for the subject's own consumption and are definitely better kept that way.

But let's consider specifically those aspects of mental reality that a person wants to communicate, to *transfer* to a fellow sentient being. Those moments of consciousness that we are most eager to share are so intimate, complex, and multifaceted as to make language often feel outright inadequate. Sometime we *wish* to be able to express our love, anger, passion, frustration, anxiety, desire, or joy, and yet we only manage to smirk, to utter "I'm so happy," to type "~!@#$%^&*," or just to burst into laughter or tears. And emotional feelings are not necessarily the most difficult mental states to inform others of. How often do teachers repeat complicated concepts to their students on average, or parents to their children, before their intended audience "gets it"? How tricky is it for patients to describe important physical symptoms to their physicians, how delicate for witnesses to convey their testimonies precisely and unambiguously, how demanding for project managers to explain lengthy details to their teams?!

In the most worthwhile and meaningful moments of our lives, in the very core of any and all human relationships, throughout all nations, cultures, and ages, *communication* (not health, wealth, or technology) is the ultimate barrier.

We can therefore reassess the implications of a real-time neuron-level brain interface with the capability of measuring and alter the activity of the circuit. If we know how to map these data to mental states and vice versa, we could instantiate the most effective and complete form of communication between human beings: the exchange of intended aspects of experience directly (and quasi-instantaneously) as opposed to through language.[12]

Imagine a community without misunderstandings, biased interpretations, partial accounts, messages lost in translation, and unintentional incomprehension. Such changes would have effects of epic proportions on the very fabric of human society and even the nature of the human spirit. We can contemplate whether mankind would rise to new heights in such a different world or rapidly edge toward self-destruction. But it is hard to imagine that mapping the roots of the mind onto the trees of the brain would be inconsequential.

Epilogue

Our journey through the brain forest projected us from the tightly packed branches of nerve cells to future civilizations founded on direct neural encoding of experiential communication. It's now time to return to the present and to recognize the many challenges that remain in the path to deciphering the mysterious divide between nervous systems and mental states.

In this book I maintained that the geometry of neuronal trees constitutes a fundamental determinant of both the instantaneous state of the mind and its continuous temporal flow and evolution. The content of all aspects of mental experience is mediated by a massively distributed pattern of electrical activity zipping through an immense number of neuronal branches extending for millions of miles, yet all unbelievably packed within the skull. The trillions of contacts among axonal and dendritic trees, dictating how signals can or cannot travel throughout the network, establish which mental states a given brain may instantiate at any one time, much as the grid of streets in a city constrains the possible traffic patterns. Perhaps most strikingly, the shape, position, and orientation of every neuronal arbor also determine the subset of neuronal pairs that *could* connect. This principle defines in return the aspects of experience a given brain is equipped to learn. On overlapping fibers, just a twitch of a synapse changes learning capability into knowledge.

If these theses are correct, then a deep understanding of the mind-brain relationship must pass through a comprehensive mapping of the structure, activity, and plasticity of all these mighty trees. We can noninvasively image the structure and activity of the aware human brain and even temporarily alter its dynamics, but the spatial and temporal scales of these approaches are completely off relative to the size and rhythms of neuronal trees. We examined the many obstacles lying in the way of completing these whole-brain neural-level maps: the astronomical number of neurons in a single brain, the minuscule caliber of their branches below the resolution limit of light, their macroscopic spatial extent spanning across brain hemispheres, and their thick mingling in one continuous, space-filling neural tissue. Until recently, neuroscientists could only examine neurons in dead brain slices, only a few at a time, and with incompletely stained arbors. Recent breakthroughs are dismantling these experimental limitations one by one, and today neuroscience labs are populated with transparent brains, genetically multicolored axons and dendrites, and neurons that literally light up when firing a spike in live behaving animals. The path to travel is still quite long, but there is reason for optimism.

And yet the neurotechnological ordeal to map every branch in all the neurons of a brain and to record their voltage in real time is just a start. Once we build the all-powerful, noninvasive, whole-brain microscanner, data will start flowing in our computers. Those hundreds of billions of neurons, branches, synapses, spikes, and synaptic signals will need to be organized, understood, taken apart, and assembled back together, not in a static picture but in the dynamic flux they constitute. Computer science has yet to devise databases and knowledge bases of such ambitious scope to deal with the breathtaking flood of information expected from the analysis of even a single day in the life of a brain. In comparison, sequencing the genome of all species on Earth, charting the course of each visible celestial body, and cataloguing every last subatomic particle all pale relative to the quest of characterizing the neural basis of experience. It will take generations of highly trained scientists, powerful machines, and tremendous societal coordination to map it all.

Even the most exuberant success in acquiring and understanding all relevant neural data, however, will not suffice without equivalent progress on the mind side of the equation. Any human being, every conscious soul, by definition *knows* what experience feels like. But self-evident as it may be, the private, intimate awareness of individuals doesn't qualify as scientific. To be counted as evidence in the study of consciousness, these observations must be communicated unambiguously, reliably, and precisely to independent observers. We need to develop a scientific approach to the mind that enables quantitative descriptions of mental states. Computational linguistics may offer a first crude approximation, but ultimately we'll need to

Epilogue

invent a way to express mental states directly. A promising direction in this regard is offered by semantic mapping, an approach to characterize cognitive spaces geometrically. It seems fair to concede, however, that such a topic belongs to a different book.

Early in the book we pondered the ultimate question of the very nature of Reality. We considered the intuitive appeal of mainstream philosophical positions such as Materialism, Idealism, and Dualism, as well as their logical shortcomings. Our understanding of the architecture and dynamics of the brain forestry is admittedly far too limited to provide a satisfactory solution to the mind-brain problem. Nonetheless, we hope to have stimulated a sufficient appreciation for the information processing power of neuronal arbors to allow a glimpse into its crucial role in explaining mental phenomena.

Based on the principles exposed in this book, we can attempt to revisit the question of Reality. In their bare computational essence, brains can be viewed as gigantic networks whose sets of connections represent associations of observables learned through experience. Note that in this framework experience can equally represent the spontaneous thought of an abstract idea or the direct perception of a physical stimulus.

We can thus offer a radical perspective of Reality. Reality constitutes an enormous interconnected web of co-occurring events. Each pair of events can be quantitatively expressed in the context of the entire web by the conditional probability representing the information content of their co-occurrence. Every human being (as well as, of course, all other animals and inanimate objects) is immersed in this universal web. From within, each person at any given moment witnesses a small fraction of co-occurring events based on his or her location, time, state of attention, and so forth. Brains evolved as networks (of neurons) themselves in order to represent most effectively the surrounding reality, thereby gaining predictive power and endowing their carriers with survival fitness. Such a view of Reality needs not be taken literally. Instead, it can simply be entertained as a convenient formalism to represent material and mental events alike. Such integration provides a natural conceptual framework to characterize the interactions among the world, the brain, and the mind.

My most sincere aspiration with this book is to have succeeded in sharing my sense of awe for those branching filaments making up our nervous systems. I've always been a nature lover, and as a child I often asked rhetorically, Who wouldn't like trees? When I started studying neuroscience I gradually realized the biological and computational implications of the arboreal structure of neurons. While thinking of neuronal circuitry, my mind's eyes saw gardens, woods, and forests. Now, when jogging in the nearby park, as soon as I let my mind wander through the surrounding vegetation, I find myself immersed in a world of axons, dendrites, and synapses. And from the joy of contemplation, I come to muse, Who wouldn't like neurons?

Notes

Preface

1. To begin with, a basic introduction to the brain and its cellular organization is provided gratis by "Brain Facts" published by the Society for Neuroscience (http://brainfacts.org). That entire book can be freely downloaded as a pdf or as an audio file. Alternatively, individual topics can be searched and browsed online. Other useful online resources including reviews of many neuroscience books can be found at http://faculty.washington.edu/chudler/neurok.html

Chapter 1

1. Rosenberg AJ (1993) The Book of Genesis—A New English Translation, Vol. 1. Judaica Press. Accessed at http://www.chabad.org/library/bible_cdo/aid/8166.

2. Rosenberg AJ (1993) The Book of Genesis—A New English Translation, Vol. 1. Judaica Press. Accessed at http://www.chabad.org/library/bible_cdo/aid/8167/jewish/Chapter-3.htm.

3. http://lhc.web.cern.ch/lhc.

4. ATLAS Consortium (2012) A particle consistent with the Higgs boson observed with the ATLAS detector at the Large Hadron Collider. Science 338(6114):1576–82.

5. http://wlcg.web.cern.ch.

6. … or in a simulation? The "simulation hypothesis" argues that if one believes that sooner or later consciousness will be simulated in machines, the chance that *we* are not living in one such computational model is very small (http://www.simulation-argument.com). Versions of this idea date back to ancient philosophy and have been proposed in many recent sci-fi books and movies. Nick Bostrom, who formally introduced this position to the modern scholarly discourse, recently published a comprehensive book: Bostrom N (2014) Superintelligence—Paths, Dangers, Strategies. Oxford: Oxford University Press.

7. For an accessible introduction to this topic, see Chalmers DJ (1995) The puzzle of conscious experience. Sci Am. 273:80–86.

8. A bold attempt in this direction is found in Tononi G (2012) Phi: A Voyage from the Brain to the Soul. New York: Pantheon.

9. Ascoli GA (2013) The mind-brain relationship as a mathematical problem. ISRN Neuroscience. 2013:1–13. http://dx.doi.org/10.1155/2013/261364

10. Aristotle (350 B.C.E) Politics. Translated by Benjamin Jowett. From The Internet Classics Archive by Daniel C. Stevenson, Web Atomics. http://classics.mit.edu/Aristotle/politics.mb.txt

11. Eyre MD, Antal M, Nusser Z (2008) Distinct deep short-axon cell subtypes of the main olfactory bulb provide novel intrabulbar and extrabulbar GABAergic connections. J Neurosci. 28(33):8217–29.

12. These hypothetical beings are referred to in modern philosophy as *zombies*; see, e.g., Kirk R (2011) Zombies. In Zalta EN (ed). Stanford Encyclopedia of Philosophy (Spring 2011 Edition). www.plato.stanford.edu/archives/spr2011/entries/zombies.

13. See also Ascoli GA, Grafman J, eds (2005) Consciousness, Mind and Brain. Milan: Massom Publisher.

14. Such a capability, initially documented in a single subject known by the initials AJ, has since been extended to several similar cases: McGaugh JL, LePort A (2014) Remembrance of all things past. Sci Am. 310(2):40–45. www.scientificamerican.com/article/the-discovery-of-super-memories.

Chapter 2

1. Parents of teenage boys might sometime suspect otherwise.

2. Helmstaedter M, Sakmann B, Feldmeyer D (2009) The relation between dendritic geometry, electrical excitability, and axonal projections of L2/3 interneurons in rat barrel cortex. Cereb Cortex. 19(4):938–50.

3. Nikolenko V, Poskanzer KE, Yuste R (2007) Two-photon photostimulation and imaging of neural circuits. Nat Methods. 4(11):943–50.

4. Goldberg JH, Tamas G, Aronov D, Yuste R (2003) Calcium microdomains in aspiny dendrites. Neuron. 40(4):807–21.

5. Barnard ES (2002) New York City Trees: A Field Guide for the Metropolitan Area. New York: Columbia University Press.

6. To help your imagination, see centralparknature.com for an actual map of most of Central Park trees!

7. According to an estimate of the Wisconsin County Forests Association, "a mature, healthy tree can have 200,000 leaves."

8. See also Ascoli GA, ed (2002) Computational Neuroanatomy: Principles and Methods. Totowa, NJ: Humana Press.

9. The Georgia Forestry Commission reports the following mature dimensions for representative medium-size trees: live oak (*Quercus virginiana*), 40–50 feet high and 3–4 feet in diameter; pin oak (*Quercus palustris*), 70–80 feet high and 2–3 feet in diameter; eastern red cedar (*Juniperus virginiana*), 50–80 feet high and 1–1.5 feet in diameter; southern red oak (*Quercus falcate*), 70–80 feet high and 2–3 feet in diameter; and water oak (*Quercus nigra*), 60–100 feet high and 2–3 feet in diameter.

10. Today there are approximately 100,000 known species of trees that exist throughout the world, according to World Resources Institute. The number of neuron types is still an open question (see also section 8.4), but if I have to venture a guess, I would use that same figure as an order of magnitude.

11. Some synapses are established from axons directly onto the soma, others from axons onto other axons, and still others from dendrites onto dendrites. The vast majority of synapses, however, are from axons onto dendrites.

12. Brown K, Sugihara I, Shinoda Y, Ascoli GA (2012) Digital morphometry of rat cerebellar climbing fibers reveals distinct branch and bouton types. J Neurosci. 32:14670–84.

13. A small proportion of synapses, usually dendrodendritic, are in fact electric, called *gap junctions*.

14. For a beautifully written journey on this topic, see Alkon DL (1992) Memory's Voice: Deciphering the Brain-Mind Code. New York: Harper Collins.

15. To be precise, the most recent scholarly estimate is 89 billion: see Herculano-Houzel S (2009) The human brain in numbers: A linearly scaled-up primate brain. Front Human Neurosci. 3:31.

16. Technically, even with a standard resolution of 600 dots per inch (dpi), a printer could fill 12,600,000 locations on a 5-inch by 7-inch image such as figure 2.5 (35×6002). At maximum contrast (50% fill, that is alternating black and white), as many as 6,300,000 dots could then be printed, thirty-five times as many as those actually present in figure 2.5. However, they would be invisible to the naked eye even by very close inspection.

17. According to their website, the 2014 collection of the Library of Congress included more than 36.8 million cataloged books and other print materials in 470 languages. http://www.loc.gov/about/general-information/.

18. Google estimated in 2010 (as part of their "Google Books" project) that nearly 130 million distinct books can be uniquely identified. http://booksearch.blogspot.com/2010/08/books-of-world-stand-up-and-be-counted.html.

19. In 2008 NPR reported that data from NASA satellites suggested some 400 billion trees on earth (~60 per person). I suspect the number of trees may have declined since then (while the number of people definitely increased).

Chapter 3

1. These channels are called *voltage-gated*. The voltage, named after Alessandro Volta, the inventor of the battery, is the difference in electric potential per unit charge between two points, that is, electric tension.

2. This transition region is also known as *axonal hillock*.

3. In neurons, this is called the *refractory period*.

4. These values should not be construed to support the urban myth that we use less than 10% of our brain. The vast majority of neurons fire *eventually*, they just don't fire all together.

5. Due to the exact same mechanism, at branch points in the axonal tree, outgoing spikes "double-up" and continue down both sides of the bifurcation, each identical to the "parent" spike.

6. A comparison between brains and computers is beyond the scope of this book and indeed is the topic of several other books. A web search of "compare brain computer" pulls some 80 million records, the first several hundreds of which appear to contain relevant and useful discussions. For a brief and referenced scholarly discussion, see Nagarajan N, Stevens CF (2008) How does the speed of thought compare for brains and digital computers? Curr Biol. 18:R756–58.

7. The information content reflected in the overall spike frequency is called *rate code*, whereas the distinction related to the exact timing of spikes within a pattern is called *time code*.

8. Quilichini P, Sirota A, Buzsáki G (2010) Intrinsic circuit organization and theta-gamma oscillation dynamics in the entorhinal cortex of the rat. J Neurosci. 30(33): 11128–42.

9. Tukker JJ, Lasztóczi B, Katona L, Roberts JD, Pissadaki EK, Dalezios Y, Márton L, Zhang L, Klausberger T, Somogyi P (2013) Distinct dendritic arborization and in vivo firing patterns of parvalbumin-expressing basket cells in the hippocampal area CA3. J Neurosci. 33(16):6809–25.

10. Packer AM, Yuste R (2011) Dense, unspecific connectivity of neocortical parvalbumin-positive interneurons: A canonical microcircuit for inhibition? J Neurosci. 31(37):13260–71.

11. These channels are called *ligand-gated*. The ligand (Latin for "to be bound") is the molecule that binds to the channels, that is, the neurotransmitter.

12. The term GABA is capitalized because it is an acronym, standing for "*g*amma-*a*mino*b*utano*ic a*cid."

13. Because the excess negative charge inside exerts a repulsive action on chloride, the effect of GABA is not symmetric and opposite to that of glutamate but, rather, typically slower and in some cases weaker.

14. Sulkowski MJ, Iyer SC, Kurosawa MS, Iyer EP, Cox DN (2011) Turtle functions downstream of Cut in differentially regulating class specific dendrite morphogenesis in *Drosophila*. PLoS One. 6(7):e22611.

15. Ropireddy D, Scorcioni R, Lasher B, Buzsáki G, Ascoli GA (2011) Axonal morphometry of hippocampal pyramidal neurons semi-automatically reconstructed after in vivo labeling in different CA3 locations. Brain Struct Funct. 216(1):1–15.

16. For a recent scholarly review, see Silver RA (2010) Neuronal arithmetic. Nat Rev Neurosci. 11(7):474–89.

17. Ferrante M, Migliore M, Ascoli GA (2013) Functional impact of dendritic branch-point morphology. J Neurosci. 33(5):2156-65.

18. Hebb DO (1949) The Organization of Behavior: A Neuropsychological Theory. New York: Wiley.

19. For a scholarly review of this classic view, see Tsien JZ (2000) Linking Hebb's coincidence-detection to memory formation. Curr Opin Neurobiol. 10:266–73.

20. This is obviously a much oversimplified cartoon. For an in-depth scholarly discussion, see Wasserman EA, Miller RR (1997) What's elementary about associative learning? Annu Rev Psychol. 48:573–607.

21. Kawaguchi Y, Karube F, Kubota Y (2006) Dendritic branch typing and spine expression patterns in cortical nonpyramidal cells. Cereb Cortex. 16(5):696–711.

22. The timberline is the mountain altitude above which trees are no longer found for lack of oxygen and freezing temperature. On the Alps, the timberline is at approximately 6000 feet above sea level.

23. The axons of only a few principal neurons of the mammalian cerebral cortex have been traced in a sufficiently extended fashion to be considered at least putatively representative of a complete axonal arbor. For those, the full length typically ranges between 0.5 meter and 1 meter. The whole dendritic arborization for those

same neurons typically sums up to 10–20 millimeters in total length, that is, less than 2% relative to the axon. Glial cells are believed to be about as abundant as neurons, but their branching is no more complex than that of dendrites. Veins, capillaries, and arteries, comparatively speaking, contribute a negligible length.

24. One hundred billion neurons times ~0.8 meters of estimated axonal length per neuron.

25. The US Central Intelligence Agency (CIA) posts the World country ranking by road network size (https://www.cia.gov/library/publications/the-world-factbook/rankorder/rawdata_2085.txt). In 2014 the top three were (total length in miles): United States 4,092,842; India 2,914,212; and China 2,551,660.

26. Specifically, the average diameter of an axonal branch is approximately one-thirtieth of a micrometer. The entire volume of all axonal branches in a brain can be estimated, approximating the cable to a cylinder, as just about one quart, which is about 60% of the whole brain size.

Chapter 4

1. See also Ikegaya Y, Aaron G, Cossart R, Aronov D, Lampl I, Ferster D, Yuste R (2004) Synfire chains and cortical songs: Temporal modules of cortical activity. Science. 304(5670):559–64.

2. For a lay overview, see Quiroga RQ, Fried I, Koch C (2013) Brain cells for grandmother. Sci Am. 308(2):30–35. For a comprehensive review, see Quiroga RQ (2012) Concept cells: The building blocks of declarative memory functions. Nat Rev Neurosci. 13(8):587–97.

3. Henze DA, Cameron WE, Barrionuevo G (1996) Dendritic morphology and its effects on the amplitude and rise-time of synaptic signals in hippocampal CA3 pyramidal cells. J Comp Neurol. 369(3):331–44.

4. Scorcioni R, Ascoli GA (2005) Algorithmic reconstruction of complete axonal arborizations in rat hippocampal neurons. Neurocomputing. 65–66:15–22.

5. Carnevale NT, Tsai KY, Claiborne BJ, Brown TH (1997) Comparative electrotonic analysis of three classes of rat hippocampal neurons. J Neurophysiol. 78(2):703–20.

6. In laboratory rodents it is possible to measure the spiking activity of hippocampal neurons while the animal roams around the cage or a maze. Researchers typically find distinct place cells that fire next to the entrance of the cage, near the water spout, or in the far corner. When the firing patterns are recorded from a sufficiently large number of neurons, each location is systematically found to be represented by multiple cells (see nobelprize.org/nobel_prizes/medicine/laureates/2014/advanced-medicineprize2014.pdf).

7. The first contribution was the synaptic plasticity rule described in section 3.3.

8. This requirement may not apply to the great deal of brain activity that is not consciously accessible.

9. This corollary does not refer to physically impossible mental states such as seeing a color outside of the visible range or moving one's muscles at the speed of light. I am also not pointing to the fact that the vast majority of all the knowledge that human beings are in principle capable of apprehending has yet to be discovered.

10. Even non–violin players can arguably "imagine" *playing*, but here we refer to the inability to imagine the actual *feelings* experienced by violin players.

11. See http://connectomethebook.com/.

12. See http://neuroscienceblueprint.nih.gov/connectome/.

13. A clear demonstration is provided by the ever-improving functionality of brain-controlled robotic arms, for example, see Hochberg LR, Bacher D, Jarosiewicz B, Masse NY, Simeral JD, Vogel J, Haddadin S, Liu J, Cash SS, van der Smagt P, Donoghue JP (2012) Reach and grasp by people with tetraplegia using a neurally controlled robotic arm. Nature. 485(7398):372–75.

14. Vetter P, Roth A, Häusser M (2001) Propagation of action potentials in dendrites depends on dendritic morphology. J Neurophysiol. 85(2):926–37.

15. In the scientific discourse, these observables are sometime called *epiphenomena* (singular: epiphenomenon).

16. Buzsaki G. (2006) Rhythms of the Brain. Oxford: Oxford University Press.

17. This argument supports the notion that mental states depend on information processing, a notion embraced most notably by Giulio Tononi in his "Information Integration Theory" of consciousness (see also note 8 in chapter 1). Information is measured in "bits," binary units that are typically expressed as either 0 or 1 (or true/false, on/off, etc.). For an accessible explanation, see Carl Zimmer's September 20, 2010 New York Times article "Sizing up Consciousness by Its Bits" (http://www.nytimes.com/2010/09/21/science/21consciousness.html) and references therein.

18. Trevelyan AJ, Sussillo D, Watson BO, Yuste R (2006) Modular propagation of epileptiform activity: Evidence for an inhibitory veto in neocortex. J Neurosci. 26(48):12447–55.

19. Chiang PH, Wu PY, Kuo TW, Liu YC, Chan CF, Chien TC, Cheng JK, Huang YY, Chiu CD, Lien CC (2012) GABA is depolarizing in hippocampal dentate granule cells of the adolescent and adult rats. J Neurosci. 32(1):62–67.

Chapter 5

1. Tononi G, Cirelli C (2013) Perchance to prune. During sleep, the brain weakens the connections among nerve cells, apparently conserving energy and, paradoxically, aiding memory. Sci Am. 309(2):34–39. A video of a scientific seminar by the same author can be found at http://www.scientificamerican.com/article/sleep-brains-way-staying-balance-video-giulio-tononi. For a more in-depth scholarly review, see Tononi G, Cirelli C (2014) Sleep and the price of plasticity: From synaptic and cellular homeostasis to memory consolidation and integration. Neuron. 81(1):12–34. There are many other theories aiming to explain the phenomenon of memory consolidation during sleep. However, this fascinating but still hotly debated topic is outside the scope of this book.

2. Edelman G (1987) Neural Darwinism: The Theory of Neuronal Group Selection. New York: Basic Books.

3. For an image of the dendritic arbors of these neurons, turn a few pages to figure 7.2.

4. Babies born blind in one eye, or losing an eye early in life grow up without ocular dominance stripes in the cortex: the "good" eye rapidly takes over all of available connections in the visual cortex. In contrast, adult-onset monocular deprivation leaves the ocular dominance stripes in the cortex largely intact for several years post-lesion.

5. See, for example, http://www.ncbi.nlm.nih.gov/books/NBK11007 or the original report: Johnson JS, Newport EL (1989) Critical period effects in second language learning: The influence of maturational state on the acquisition of English as a second language. Cogn Psychol. 21(1):60–99.

6. For example, finding out that your spouse has had an affair or accepting "impossible" mathematical entities such as the square root of negative numbers.

7. Budd JM, Kovács K, Ferecskó AS, Buzás P, Eysel UT, Kisvárday ZF (2010) Neocortical axon arbors trade-off material and conduction delay conservation. PLoS Comput Biol. 6(3):e1000711.

8. For scholarly reviews in human and animal models, see respectively: May A (2011) Experience-dependent structural plasticity in the adult human brain. Trends Cogn Sci. 15(10):475–82; and Caroni P, Donato F, Muller D (2012) Structural plasticity upon learning: Regulation and functions. Nat Rev Neurosci. 13(7):478–90.

9. The capability of experiencing a mental state, that is, knowledge, should not be confused with the capability of learning. The capability of learning does *depend* on preexisting knowledge, as explained in the next chapter.

10. Holtmaat A, Svoboda K (2009) Experience-dependent structural synaptic plasticity in the mammalian brain. Nat Rev Neurosci. 10(9):647–58.

11. Rihn LL, Claiborne BJ (1990) Dendritic growth and regression in rat dentate granule cells during late postnatal development. Brain Res Dev Brain Res. 54(1): 115–24.

12. Tamamaki N, Nojyo Y. (1991) Crossing fiber arrays in the rat hippocampus as demonstrated by three-dimensional reconstruction. J Comp Neurol. 303(3):435–42.

13. Golding NL, Kath WL, Spruston N (2001) Dichotomy of action-potential backpropagation in CA1 pyramidal neuron dendrites. J Neurophysiol. 86(6):2998–3010.

14. This is sometimes referred to as working memory or short-term memory. By learning, in contrast, here we refer to long-term memory storage, that is, the ability to reinstate a specific mental state at a future time well beyond the current temporal span of attention.

15. There are rare known cases of people possessing highly superior autobiographical memory. They may be considered the closest living embodiment of the hypothetical scenario of "learning everything." This condition, technically named "hyperthymestic syndrome," was recently described in both the biomedical literature and television programs for the lay public (see note 14 in chapter 1).

Chapter 6

1. This requirement was in fact explicitly stated by Hebb in his original formulation of synaptic plasticity (see section 3.3 for the complete quote and reference): "When an axon of cell A is *near enough* to excite B and repeatedly or persistently takes part in firing it…" (emphasis ours). The direction of the stimulation implies that the axon belongs to neuron A and the dendrite to neuron B.

2. Stepanyants A, Chklovskii DB (2005) Neurogeometry and potential synaptic connectivity. Trends Neurosci. 28(7):387–94.

3. Botanical trees, like all biological entities, are part of a complex interaction network of the ecosystem and do in fact communicate directly with chemicals and indirectly through other agents (e.g., insects). Here our assertion is more specific to the notion of structural networks and information-processing devices. Nevertheless, even this narrower position might be challenged by novel anthropological ideas; see Kohn E (2013) How Forests Think: Toward an Anthropology Beyond the Human. Berkeley/Oakland: University of California Press.

4. Measurements in a few cortical regions of the rodent brain found actual synapses in only a minority of axonal-dendritic overlaps, ranging from 25% to less than 10%.

We expect this fraction to vary widely across regions of the nervous system and possibly also among species and developmental stages. Nevertheless, the value of 10% is likely to represent an excellent approximation for at least some important parts of the brain. At any rate, our reasoning is not affected by this specific numerical detail.

5. This average estimate assumes that all neurons are similar and the brain is homogeneous. Subsequent chapters argue that this is emphatically *not* the case, but the simple computations in this paragraph are still valid at the level of order of magnitude (within a tenfold factor).

6. The combined chance of a dendrite of a random neuron C to overlap with the axon from neuron A *and* of the axon of the same neuron C to overlap with a dendrite of neuron B is one in a trillion. Thus, the probability that any one neuron C would *not* indirectly connect A and B by axonal-dendritic overlap is 0.999999999999 (zero followed by twelve 9's). However, the chance that neither of two random neurons (C and D) would bridge A and B is 0.999999999999 × 0.999999999999. Because there are 100 billion neurons to try, the chance that A and B fail to encounter via one stop is 0.999999999999 multiplied by itself 100 billion times. The result of this operation is approximately 90%. This means that one in ten pairs of neurons may have a "once-removed" axonal-dendritic overlap.

7. This observation is also corroborated by recent experimental evidence: Sadtler PT, Quick KM, Golub MD, Chase SM, Ryu SI, Tyler-Kabara EC, Yu BM, Batista AP (2014) Neural constraints on learning. Nature. 512(7515):423-6.

8. As explained in section 6.4, continuous stimulation can induce formation of new axonal-dendritic overlaps by lengthening their respective branches.

9. Specifically, the principle states that the probability of an axon A_1 overlapping with a dendrite D_1 depends on the number of other dendrites $D_2 \ldots D_n$ receiving connections from A_1 and on the number of other axons $A_2, \ldots A_m$ that make synapses onto $D_1, \ldots D_m$.

10. Shepherd GM, Svoboda K (2005) Laminar and columnar organization of ascending excitatory projections to layer 2/3 pyramidal neurons in rat barrel cortex. J Neurosci. 25(24):5670–79.

11. The orders of magnitude for the size of a thin axonal branch, an ionic channel, and an ion are respectively tens of nanometers, nanometers, and tenths of nanometers (10^{-8}, 10^{-9}, and 10^{-10} meters), respectively. See also http://bionumbers.hms.harvard.edu.

12. Sebastian Seung (see note 11 in chapter 4) similarly distinguishes the "4 R" mechanisms to alter neural circuits: (1) Reweighting or change of synaptic strength; (2) Reconnection or creation/elimination of synapses; (3) Rewiring or movement of neural branches; and (4) Regeneration or death/birth of neurons.

13. This assumption is consistent with the principle of "population coding," which we discuss more directly in section 8.4.

Chapter 7

1. This is never the case even for identical twins because they have separate embryonic development and their neuronal differentiations involve independent stochastic molecular processes. More details about the developmental process are explained in the next section.

2. Noninvasive imaging ("brain scans") can visualize the coarse structure of the human nervous system but lacks the resolution required to detect fine structures such as dendrites and axons. In order to visualize these branching structures, a dye needs to be inserted in the cell body of an individual neuron. Sufficient time must then elapse for the dye to reach all parts of the arbor. In the case of extensive axons the necessary time for the dye to fill the entire structure can be several weeks. During such a long period, tissue degeneration blocks the movement of the dye. The specific technical reasons are somewhat more subtle than described here, but the fact remains that the full axonal extent of human brains cannot be visualized with current technology. See also section 9.3 for additional discussion.

3. Apes (especially chimpanzees and bonobos) are even closer to humans, but they are just too close for modern society to accept the ethical burden of carrying out invasive research on them. Recently research on monkeys has also seen a drastic reduction for similar considerations.

4. Harris TW, Antoshechkin I, Bieri T, Blasiar D, Chan J, Chen WJ, De La Cruz N, Davis P, Duesbury M, Fang R, Fernandes J, Han M, Kishore R, Lee R, Müller HM, Nakamura C, Ozersky P, Petcherski A, Rangarajan A, Rogers A, Schindelman G, Schwarz EM, Tuli MA, Van Auken K, Wang D, Wang X, Williams G, Yook K, Durbin R, Stein LD, Spieth J, Sternberg PW (2010) WormBase: A comprehensive resource for nematode research. Nucleic Acids Res. 38(Database issue):D463–67.

5. Jacobs B, Lubs J, Hannan M, Anderson K, Butti C, Sherwood CC, Hof PR, Manger PR (2011) Neuronal morphology in the African elephant (*Loxodonta africana*) neocortex. Brain Struct Funct. 215(3–4):273–98.

6. Kubota Y, Shigematsu N, Karube F, Sekigawa A, Kato S, Yamaguchi N, Hirai Y, Morishima M, Kawaguchi Y (2011) Selective coexpression of multiple chemical markers defines discrete populations of neocortical GABAergic neurons. Cereb Cortex. 21(8):1803–17.

7. Oginsky MF, Rodgers EW, Clark MC, Simmons R, Krenz WD, Baro DJ (2010) D(2) receptors receive paracrine neurotransmission and are consistently targeted to a subset of synaptic structures in an identified neuron of the crustacean stomatogastric nervous system. J Comp Neurol. 518(3):255–76.

8. Rodger J, Drummond ES, Hellström M, Robertson D, Harvey AR (2012) Long-term gene therapy causes transgene-specific changes in the morphology of regenerating retinal ganglion cells. PLoS One. 7(2):e31061.

9. The most direct approach for measuring the electric activity of neurons is to insert microelectrodes into the nervous system to record voltage. Nevertheless, advanced molecular techniques have also been developed to visually detect electric changes by optical microscopy utilizing dyes that transiently change color as a result of neuronal firing (see section 9.3).

10. Most individuals in this species are hermaphrodites. In the small minority of males, the number of neurons is 383.

11. In truth, the exact extent of such "stereotypy" is not yet completely known because not enough specimens have been investigated down to every last neuron. The entire nervous system of only one (male) worm has been mapped neuron by neuron, and those of two more animals have been mapped partially. Stereotypy has been (incompletely) determined by looking at individual neurons in several dozen animals.

12. For a free scholarly overview, see Stiles J, Jernigan TL (2010) The basics of brain development. Neuropsychol Rev. 20(4):327–48. Available at ncbi.nlm.nih.gov/pmc/articles/PMC2989000.

13. This average retraction is only relative to overall brain size. In absolute terms, dendrites and especially axons tend to expand together with the head, skull, and brain, thus yielding a net elongation of the cable. Brain size quadruples in the first five years of the baby's life, but the adult brain volume is only 10% larger than that matured by age six. The number of axonal-dendritic overlaps, however, depends on the spatial density of the trees rather than their absolute length. The brain volume grows with the third power of the radius, causing a steep reduction of axonal and dendritic densities.

14. For example, although neuronal generation and migration have already largely ended at birth, the proliferation, differentiation, and positioning of glial cells and of their progenitors mostly occur in infancy and through childhood. Glia could therefore be responsible for neuronal plasticity during this period.

15. See www.translatingtime.net.

16. See also Greenwood PM, Parasuraman R (2012) Nurturing the Older Brain and Mind. Cambridge, MA: MIT Press.

17. Scorza CA, Araujo BH, Leite LA, Torres LB, Otalora LF, Oliveira MS, Garrido-Sanabria ER, Cavalheiro EA (2011) Morphological and electrophysiological properties of pyramidal-like neurons in the stratum oriens of Cornu ammonis 1 and Cornu ammonis 2 area of *Proechimys*. Neuroscience. 177:252–68.

18. Pyapali GK, Turner DA (1996) Increased dendritic extent in hippocampal CA1 neurons from aged F344 rats. Neurobiol Aging. 17(4):601–11.

19. Wang X, Kim JH, Bazzi M, Robinson S, Collins CA, Ye B (2013) Bimodal control of dendritic and axonal growth by the dual leucine zipper kinase pathway. PLoS Biol. 11(6):e1001572.

20. Borst A, Haag J (1996) The intrinsic electrophysiological characteristics of fly lobula plate tangential cells: I. Passive membrane properties. J Comput Neurosci. 3(4):313–36.

21. Li Y, Brewer D, Burke RE, Ascoli GA (2005) Developmental changes in spinal motoneuron dendrites in neonatal mice. J Comp Neurol. 483:304–17.

22. Lu J, Tapia JC, White OL, Lichtman JW (2009) The interscutularis muscle connectome. PLoS Biol. 7(2):e32.

23. Wu H, Williams J, Nathans J (2012) Morphologic diversity of cutaneous sensory afferents revealed by genetically directed sparse labeling. Elife. 1:e00181.

24. Cerebellar granule cells are not to be confused with homonymous granule cells in other brain regions, such as hippocampal granule cells and olfactory granule cells. The naming ambiguity in this case dates back to when only neuronal somata were visible under the microscope, and these three distinct neuron types all happened to have granule-shaped cell bodies! Their axonal and dendritic arbors, however, are completely different in each of the three types.

25. See http://neurolex.org/wiki/Category:Neuron.

26. See http://BrainInfo.org.

27. Shepherd GM, Grillner S (2010) Handbook of Brain Microcircuits. Oxford: Oxford University Press.

28. The terms drivers and modulators was originally introduced in the visual system to distinguish neurons transmitting information about the content of the field of view from those altering the probability of that transmission: Sherman SM, Guillery RW (1998) On the actions that one nerve cell can have on another: Distinguishing "drivers" from "modulators." Proc Natl Acad Sci USA. 95(12):7121–26. However, *these* modulators should not be confused with modulatory neurotransmitters and the process of synaptic neuromodulation briefly mentioned at the end of section 3.3.

29. Staiger JF, Flagmeyer I, Schubert D, Zilles K, Kötter R, Luhmann HJ (2004) Functional diversity of layer IV spiny neurons in rat somatosensory cortex: Quantitative morphology of electrophysiologically characterized and biocytin labeled cells. Cereb Cortex. 14(6):690–701.

30. Ascoli GA, Brown K, Calixto E, Card P, Barrionuevo G (2009) Quantitative morphometry of electrophysiologically identified CA3b interneurons reveals robust local geometry and distinct cell classes. J Comp Neurol. 515:677–95.

31. These neuromodulatory systems were mentioned at the end of section 3.3, and their special receptors are called metabotropic as opposed to ionotropic. For completeness we should note that there are also metabotropic GABA and glutamate receptors in addition to the ionotropic receptors for these neurotransmitters, but this additional detail is unnecessary to our exposition.

Chapter 8

1. See also: DeFelipe J, López-Cruz PL, Benavides-Piccione R, Bielza C, Larrañaga P, Anderson S, Burkhalter A, Cauli B, Fairén A, Feldmeyer D, Fishell G, Fitzpatrick D, Freund TF, González-Burgos G, Hestrin S, Hill S, Hof PR, Huang J, Jones EG, Kawaguchi Y, Kisvárday Z, Kubota Y, Lewis DA, Marín O, Markram H, McBain CJ, Meyer HS, Monyer H, Nelson SB, Rockland K, Rossier J, Rubenstein JL, Rudy B, Scanziani M, Shepherd GM, Sherwood CC, Staiger JF, Tamás G, Thomson A, Wang Y, Yuste R, Ascoli GA (2013) New insights into the classification and nomenclature of cortical GABAergic interneurons. Nat Rev Neurosci. 14(3):202–16.

2. Hirsch JA, Martinez LM, Alonso JM, Desai K, Pillai C, Pierre C (2002) Synaptic physiology of the flow of information in the cat's visual cortex in vivo. J Physiol. 540(Pt 1):335–50.

3. Wang Y, Rubel EW (2012) In vivo reversible regulation of dendritic patterning by afferent input in bipolar auditory neurons. J Neurosci. 32(33):11495–504.

4. http://www.app.pan.pl/archive/published/app57/app20110019.pdf.

5. Glickfeld LL, Scanziani M (2006) Distinct timing in the activity of cannabinoid-sensitive and cannabinoid-insensitive basket cells. Nat Neurosci. 9(6):807–15.

6. For a scholarly account, see Petilla Interneuron Nomenclature Group (2008) Petilla terminology: Nomenclature of features of GABAergic interneurons of the cerebral cortex. Nat Rev Neurosci. 9(7):557–68.

7. Underwood E (2014) Brain's GPS finds top honor. Science 346(6206):149. See also "The Nobel Prize in Physiology or Medicine 2014" at nobelprize.org/nobel_prizes/medicine/laureates/2014/.

8. For an evolving knowledge base of hippocampal neuron types, see http://hippocampome.org.

9. For an online pointer to the NIH Human Connectome Project, see note 12 in chapter 4.

Notes

10. According to Boeing itself, the 747 also boasts 170 miles of wiring (presumably including the four engines). Yet this cable length pales, in comparison to the axonal wiring in a single *mouse* brain by a full order of magnitude (tenfold).

Chapter 9

1. MacLean JN, Watson BO, Aaron GB, Yuste R (2005) Internal dynamics determine the cortical response to thalamic stimulation. Neuron. 48(5):811–23.

2. Brains, however, do not operate in isolation from their embodiments. If Lisa and Monica exchanged their brains (or, should we say, their bodies), the consequences for their respective minds would be hard to predict but could be destructive.

3. See, however, the discussion at the end of section 4.4. The different connectivity of distinct brains implies that different spatial-temporal activity patterns may be encoding for the same concepts. Thus, in order to capture the commonality of *meaning* across individuals, a population-averaged brain should be computed based on the information that the circuitry can process. How exactly this can be achieved, however, is not yet clear.

4. See also Asbury C (2013) Brain Imaging Technologies and Their Applications in Neuroscience. New York: Dana Foundation. Available at http://www.dana.org/news/publications/publication.aspx?id=34292.

5. For an overview of the methods to investigate the tree shape of neurons, see Parekh R, Ascoli GA (2013) Neuronal morphology goes digital: A research hub for cellular and system neuroscience. Neuron. 77(6):1017–38.

6. Donohue DE, Ascoli GA (2011) Automated reconstruction of neuronal morphology: An overview. Brain Res Rev. 67(1–2):94–102. See also http://diademchallenge.org.

7. Clery D. (2014) Light loophole wins laurels. Science 346(6207):290-1. See also "The Nobel Prize in Chemistry 2014" at nobelprize.org/nobel_prizes/chemistry/laureates/2014/.

8. See, for example, Ascoli GA (2006) Mobilizing the base of neuroscience data: The case of neuronal morphologies. Nat Rev Neurosci. 7(4):318–24.

9. https://www.humanbrainproject.eu/.

10. http://www.whitehouse.gov/share/brain-initiative.

11. This admittedly far-fetched scenario is *sufficient* to provide a full description of the mind, but it is not strictly *necessary*. Population coding by cell assemblies might allow achieving the same goal more simply by measuring a statistically representative sample of neurons, synapses, and spikes.

12. For a very recent "early" proof of concept, see Grau C, Ginhoux R, Riera A, Nguyen TL, Chauvat H, Berg M, Amengual JL, Pascual-Leone A, Ruffini G (2014) Conscious brain-to-brain communication in humans using non-invasive technologies. PLoS One. 9(8):e105225. http://dx.plos.org/10.1371/journal.pone.0105225.

Index

Action potential. *See* Spike
Activity pattern (firing pattern; neural pattern), 9, 36, 59–61, 63–66, 68, 71–73, 75, 77–78, 81, 83–86, 91, 94–96, 100, 105, 107, 114, 119–120, 125, 132, 136–137, 144, 146, 149, 163–164, 173, 181, 184–188, 190–191, 195, 198, 200, 209
Amygdala, 8, 135, 139, 165
Anion. *See* Ionic channel
Aristotle, 9
Axonal bouton. *See* Varicosity
Axonal branching, 18, 30, 50, 80, 97, 168, 196–197, 201
Axonal-dendritic overlap (potential synapses), 97–98, 100, 102, 105, 107–112, 114, 116, 118–122, 125, 133–137, 149, 182, 185, 191, 201

Background information, 101–102, 105, 108–111, 116
Basal ganglia, 8, 135, 137
Basket cell, 18, 22, 37, 75, 143, 153, 155, 160–161, 163–164, 182
Berkeley, 6
Bifurcation, 43, 52–53, 62, 75, 128, 139, 142, 159
Bipolar cell, 143, 159
Boeing 747, 50, 179

Cation. *See* Ionic channel
Cell assembly, 45, 63, 91, 100, 105, 107, 110, 114, 120–122, 131, 136
Cell body. *See* Soma
Central nervous system. *See* Nervous system
Cerebellum, 8, 25, 71, 135, 139, 143, 152, 155, 157, 165, 168, 174
Chandelier cell, 22, 114, 143, 155, 160–161, 178
Channel. *See* Ionic channel
Circuitry (neural circuit), 20, 66, 72, 78, 80, 83–84, 89–91, 96–97, 104–105, 107–108, 112, 114, 119–120, 128, 131–132, 134, 136–137, 140, 145, 149, 161, 166–168, 170, 172–173, 181, 184–190, 200, 205, 207, 212
Concentration gradient. *See* Gradient
Connectome, 67–69, 71, 77, 87, 94, 96, 200
Consciousness, 12–15, 60, 63, 66, 72, 84, 86, 92, 181, 188–189, 207, 210
Critical period, 80–81, 135

Dendritic branching, 24, 31, 40, 44, 54, 69, 97–98, 100, 116, 118, 121, 160, 197
Dendritic spines. *See* Spine
Dentate gyrus, 75, 90, 166–168, 178
Distributed representation. *See* Population coding

EEG. *See* Electroencephalography
Electric gradient. *See* Gradient
Electroencephalography (EEG), 73, 162, 192–195
Entorhinal cortex, 37, 148, 166–168, 170, 178
EPSP. *See* Synaptic signal
Excitation, 26, 38–40, 45, 47, 49, 64, 65, 73, 144, 146–147, 149–150, 152, 157, 161–163, 165, 167–168, 170, 173, 195–196, 205
Experience, 4, 12–15, 27, 32, 36, 45, 61, 63–67, 77–79, 83–87, 91–92, 94, 96–97, 102, 104, 107–112, 114, 116, 118–121, 125, 132, 134–136, 144, 162, 165–167, 182, 184–185, 188–190, 195, 201, 206–207, 209–211

Firing pattern. *See* Activity pattern

GABA, 38–39, 147, 149–150, 152, 163, 196
Ganglion cell, 22, 128, 143
Gene (optogenetics), 6, 17, 49, 126, 130–133, 136, 163, 187–189, 195–205, 210
Glia, 9, 17, 50
Glutamate, 38–39, 47, 147, 149–150, 152, 163, 196
Gradient (concentration gradient; electric gradient; ionic gradient), 10–11, 17, 33–35, 38–39, 192
Grandmother cell, 61, 102, 122, 130
Granule cell, 22, 90, 143, 153, 157–158, 166–168, 178–179

Hebb, 45–47, 63, 96
Heraclitus, 85
Hippocampus, 8, 37, 41, 44, 61–63, 81, 89–91, 119, 135, 137, 139, 148, 150, 155, 157–158, 165–168, 170, 174, 178–179
Homunculus, 175–177

Idealism, 3, 5, 66, 92, 94, 211
Inhibition, 26, 38–40, 49, 64–65, 73, 146–147, 149–150, 152, 157, 160–161, 163, 167–168, 170, 173, 179, 195–196
Interneuron, 10, 18, 37, 52, 71, 128, 147–150, 152–153, 156–157, 160, 167–168, 170, 173–174, 178, 182, 204
Ionic channel (anion; cation; ionic), 10, 33, 38–40, 47, 49, 140–141, 152, 162–163, 198
Ionic gradient. *See* Gradient
IPSP. *See* Synaptic signal

Kant, 5
Knowledge, 3, 15, 46, 66, 68, 71–72, 77, 85, 87, 91–92, 94, 96–97, 101–102, 105, 108, 110–112, 114. 116, 119–121, 15–126, 128, 132, 134–137, 145, 150, 174, 181–182, 186–187, 189–191, 200, 209–210

Language, 73, 80–81, 116, 134, 137, 147, 155, 161, 165, 174, 193–194, 206–207
Learning, 8, 45–47, 77–81, 83, 85, 87, 89, 91–92, 94, 96–97, 101–102, 105, 107–112, 114, 116, 118–122, 125, 128, 131–132, 134–135, 165, 181–182, 184, 189–191, 200–201, 209

Martinotti cells, 155, 163–164
Materialism, 5, 12, 66, 92, 94, 211
Memory, 14, 46–47, 73, 81, 83, 87, 89, 93–94, 97, 119, 131, 136, 165–166, 184–185, 189
Mental state (thought), 5, 8–9, 12–15, 59–61, 63–68, 71–73, 75, 77–78, 81, 83–88, 91–92, 94–96, 100, 102, 104–105, 107, 109, 112, 114, 119–122, 125, 132, 136, 144, 146,

149, 162, 165–166, 181–182, 184–186, 188–191, 193–194, 200–201, 206–207, 209–211
Microscopy, 20, 22, 44, 69, 130–131, 157, 195–198, 200–202

Neocortex, 6–7, 18, 27, 37, 71, 80, 89, 91, 132, 135, 137, 139, 143, 147, 150, 155, 158, 165–166, 168, 174, 182
Nervous system (central nervous system; peripheral nervous system), 5–6, 8–9, 12, 30, 45, 53, 55, 59–60, 66–67, 72, 80, 86–87, 89, 95, 98, 104, 112, 118, 120, 126, 128, 130–133, 136, 140–141, 143–144, 152, 177, 181, 185, 188, 191, 196, 200, 206, 209, 211
Network connectivity, 53, 72, 77, 81, 86, 132, 181, 187, 189–190
Neural circuit. *See* Circuitry
Neural pattern. *See* Activity pattern
Neuromodulation, 49, 151–152
Neuron numbers, 27, 29, 63, 72, 95, 100, 112, 130, 133, 173, 178, 196, 203, 210
Neurogliaform cell, 52, 128, 155–156
Neurotransmitter, 24, 26, 33, 36, 38–39, 45, 47, 49, 53, 141, 147, 152, 163, 181, 196–198

Optogenetics. *See* Gene

Pattern completion, 96
Pavlov, 46
Peripheral nervous system. *See* Nervous system
Plasticity, 14–15, 26, 43, 45–47, 49–50, 78, 80–81, 83, 86–87, 89, 91, 96, 109–110, 114, 119–120, 130, 134–135, 152, 162, 165, 181, 186, 194–195, 200–201, 205–206, 209

Plato, 5
Population coding (distributed representation), 9, 63, 175, 177
Potential synapses. *See* Axonal-dendritic overlap
Principal cell, 50, 139, 143, 146–147, 149–150, 152, 166–168, 173–175, 177–178
Proteins, 17, 33, 49, 163, 196–198, 202, 204
Purkinje cell, 71, 139, 143, 152–153, 155, 160
Pyramidal cell, 18, 22, 41, 62, 71, 75, 82, 90, 107, 139–140, 143, 146–147, 149–150, 152–153, 155–158, 160–161, 166–167, 170, 178

Receptor, 8, 24, 38, 45, 49, 53, 112, 130, 140–141, 144, 149, 152, 163, 165, 197–198
Retina, 8, 80, 96–97, 128, 140–141, 143–144, 157, 185, 188, 198
Rhythms, 8, 49, 73, 146, 162, 179, 209

Soma (cell body), 17–19, 24, 33, 37–40, 43–44, 47, 50, 52–55, 71, 75, 82, 95, 114, 128, 133, 139, 142–143, 146–148, 150, 152–153, 156–158, 160–161, 163–164, 168, 170–171, 174–175, 182, 195
Spike (action potential), 34–36, 38–40, 43–44, 46–47, 49, 54–55, 59–61, 63–64, 66, 81, 84–86, 91, 94–95, 104, 107, 109, 119–120, 141, 144, 146, 149, 161–164, 171, 177, 182, 184, 186–188, 191–192, 194–195, 198, 200–205, 210
Spiking pattern. *See* Activity pattern
Spinal cord, 6, 8, 30, 34, 128, 130, 141–142, 155, 158

Spine (dendritic spines), 24, 26, 53, 78, 86–87, 98, 107, 116, 120, 136, 150, 157, 158, 160, 165, 168. *See also* Spinal cord
Stellate cell, 37, 148, 155, 157, 178
Synapse formation, 89, 91, 94, 109–110, 112, 118, 182, 186
Synapse numbers, 29, 78, 95, 98, 126, 137, 187
Synaptic signal (EPSP; IPSP), 26, 38–40, 43–47, 49, 54, 85, 152, 210
Synaptic strength, 27, 45–46, 49, 83–85, 91, 119–121, 165, 182, 186

Thought. *See* Mental state
Thalamus, 8, 80, 165
Threshold, 39, 44, 46, 54, 60, 86, 162

Varicosity (axonal bouton), 24, 26, 36, 45, 53, 78, 98, 107, 116, 120, 136, 165, 168
Vision, 8, 67, 126, 130, 140, 165, 178